Hans Meyer, John Bishop Tingle

Determination of Radicles in Carbon Compounds

Hans Meyer, John Bishop Tingle

Determination of Radicles in Carbon Compounds

ISBN/EAN: 9783337060305

Printed in Europe, USA, Canada, Australia, Japan

Cover: Foto ©berggeist007 / pixelio.de

More available books at **www.hansebooks.com**

DETERMINATION OF RADICLES

IN

CARBON COMPOUNDS.

BY

DR. H. MEYER,

Docent and Adjunct of the
Imperial and Royal German University, Prague.

AUTHORIZED TRANSLATION

BY

J. BISHOP TINGLE, PH.D., F.C.S.,

Instructor of Chemistry at the Lewis Institute,
Chicago, Ill.

FIRST EDITION.

FIRST THOUSAND.

NEW YORK:
JOHN WILEY & SONS.
LONDON: CHAPMAN & HALL, LIMITED.
1899.

AUTHOR'S PREFACE.

THIS English edition of my "Anleitung zur quantitativen Bestimmung der organischen Atomgruppen" has been prepared by Dr. J. Bishop Tingle, to whom I am greatly indebted for the care he has bestowed on it. I have endeavored to bring it into conformity with the present state of the science by various corrections and additions. It has been further improved by certain changes in arrangement which Dr. Tingle has made, and he has also added various notes. The present edition is thus a decided advancement on the German one, and I trust that in its new form it may gain many new friends whilst retaining its old ones.

Dr. HANS MEYER.

PRAGUE, October 1899.

TRANSLATOR'S PREFACE.

THE success of the German edition of Dr. Meyer's book was only one of the reasons that led to the preparation of this translation. The quantitative side of organic chemistry, apart from elementary analysis, is almost always neglected in the ordinary courses of instruction, and when the need for it arises, in the prosecution of research work for instance, it is difficult to obtain a comprehensive view of the methods which are available without undue expenditure of time. This little work supplies, for the first time, a systematic treatment of these methods which, it is hoped, may help to remove this drawback and may also encourage the introduction of some *quantitative* work into the college courses of organic preparations, since such a departure could scarcely fail to be beneficial in various ways to the student. From the translator's experience with the German edition he believes that the present one will be serviceable to instructors and senior students of organic chemistry. Considerable care has been bestowed on the proof-sheets, and it is hoped that the errors which may have escaped notice are not too glaring.

LEWIS INSTITUTE, CHICAGO, ILL.,
    October 1899.

CONTENTS.

CHAPTER I.

INTRODUCTORY. DETERMINATION OF HYDROXYL, $\overset{\text{\tiny I}}{\text{OH}}$....... PAGE 1

Introductory, 1. Determination of hydroxyl, 3. Acylation, 3. Preparation of acetyl derivatives, 5. (A) By acetyl chloride, 5. (B) By acetic anhydride, 7. (C) By glacial acetic acid, 8. Isolation of acetyl derivatives, 9. Determination of the acetyl groups, 9. (A) Hydrolytic methods, 9. (B) Additive method, 14. (C) Potassium acetate method, 14. (D) Distillation method, 15. Benzoyl derivatives, 17. (A) Preparation from benzoyl chloride, 18. (B) Preparation from benzoic anhydride, 21. Preparation of substituted benzoic acid derivatives and of phenylsulphonic chloride, 22. Acylation by means of substituted benzoic acid derivatives and of phenylsulphonic chloride, 23. Analysis of benzoyl derivatives, 24. Acylation by means of other acid radicles, 27. Alkylation of hydroxyl groups, 28. Preparation of benzyl derivatives, 28. Preparation of carbamates by means of carbamyl chloride, 29. Preparation of diphenylcarbamyl chloride, 30. Preparation of phenylcarbamic acid derivatives, 31. Preparation of phenylisocyanate, 31. Action of phenylisocyanate on hydroxyl derivatives, 31.

CHAPTER II.

DETERMINATION OF METHOXYL, $\overset{\text{\tiny I}}{\text{CH}_3\text{O}}-$, ETHOXYL, $\overset{\text{\tiny I}}{\text{C}_2\text{H}_5\text{O}}-$, AND CARBOXYL, $\overset{\text{\tiny I}}{\text{CO.OH}}$.................................. 33

Determination of methoxyl, S. Zeisel's method, 33.

vii

For non-volatile substances, 36. For volatile compounds, 38. Modified method, 39. Method for the differentiation of methoxyl and ethoxyl, 40. Determination of ethoxyl, 41. Determination of carboxyl, 41. (A) Analysis of metallic salts, 42. (B) Titration of acids, 43. (C) Etherification, 44. (D) Electrolytic conductivity of sodium salts, 46. Indirect methods for the determination of the basicity of acids, 51. (i) Carbonate method, 51. (ii) Ammonia method, 52. (iii) Hydrogen sulphide method, 52. (iv) Iodine-oxygen method, 57.

CHAPTER III.

DETERMINATION OF CARBONYL.................................. 60

Preparation of phenylhydrazones, 60. Preparation of substituted hydrazones and of parabromophenylhydrazine, 63. Indirect method, 65. Preparation of oximes, 70. Preparation of semicarbazones, 74, 77. Preparation of semicarbazide salts, 75. Preparation of amidoguanidine derivatives, 78. Paramidodimethylaniline derivatives, 80.

CHAPTER IV.

Determination of the amino group, 81. Determination of aliphatic amines (i) nitrous acid method, 81. (ii) Analysis of salts and double salts, 82. (iii) Acetylation, 83. Determination of aromatic amines: (i) Titration of the salts, 83. (ii) Preparation of diazo-derivatives: (*a*) conversion into an azo dye, 84. (*b*) Indirect method, 85. (*c*) Azoimide method, 86. (*d*) Sandmeyer-Gattermann's reaction, 87. (iii) Analysis of salts and double salts, 89. (iv) Acetylation, 89. Determination of the nitrile group, 90. Determination of the amido group, 91. Determination of the imide group: (i) Acetylation, 92. (ii) Alkylation, 93. (iii) Analysis of salts, 94. (iv) Elimination of imidogen as ammonia, 94. Determination of methyl imide, 94. Determination of ethyl imide, 99. Differentiation of the methyl imide and ethyl imide groups, 99.

CONTENTS.

CHAPTER V.

Determination of the diazo-group (A) Aliphatic diazo-compounds: (i) Titration with iodine, 100. (ii) Analysis of the iodine derivative, 101. (iii) Determination of the nitrogen in the wet way, 101. (B) Aromatic diazo-compounds. Diazonium derivatives, 103. Determination of the hydrazide group, 103. (i) By oxidation, 104. (ii) Iodometric method, 106. Determination of the nitro-group. (A) Titration method, 106. (i) Method for non-volatile compounds, 108. (ii) Modifications for volatile compounds, 108. (B) Diazo-method, 109. Determination of the iodoso- and iodoxy-groups, 109. Determination of the peroxide group, 110. The iodine number, 111. Appendix. Table of the weights of a cubic centimeter of hydrogen, 116. Tension of aqueous vapor, 118. Table for the value of $\dfrac{a}{1000-a}$, 118. Index of authors, 121. Index of subjects, 129.

ABBREVIATIONS.

The following abbreviations have been used in the bibliographical references:

Am. Chem. Journ.	American Chemical Journal.
Ann.	Liebig's Annalen der Chemie und Pharmacie.
Ann. de Ch. Ph.	Annales de Chimie et de Physique.
Arch. Pharm.	Archiv der Pharmacie.
B.	Berichte der Deutschen chemischen Gesellschaft.
Bull.	Bulletins de la Société Chimique de Paris.
C.	Chemisches Centralblatt.
Ch. R.	Chemische Revue.
Ch. Ztg.	Chemiker-Zeitung.
Ch. N.	Chemical News.
C. r.	Comptes rendus de l'Académie des sciences (Paris).
Dingl.	Dingler's polytechnisches Journal.
Gazz.	Gazzetta chimica italiana.
H.	Beilstein, Handbuch.
J.	Jahresbericht über die fortschritte der Chemie.
J. Am.	Journal of the American Chemical Society.
Journ. Chem. Soc.	Journal of the Chemical Society of London.
J. pr.	Journal für praktische Chemie.
M.	Monatshefte für Chemie.
M. & J.	V. Meyer and P. Jacobson, "Lehrbuch der organischen Chemie."
Rec.	Recueil des travaux chimiques des Pays-Bas.
S.	Seelig, "Organische Reaktionen und Reagentien."
W. Ann.	Wiedemann's Annalen der Physik und Chemie.
Z.	Zeitschrift für physikalische Chemie.
Z. An.	Zeitschrift für anorganische Chemie.
Z. anal.	Zeitschrift für analytische Chemie.
Z. ang. Ch.	Zeitschrift für angewandte Chemie.
Z. f. Ch.	Zeitschrift für Chemie.
Z. physiol. Ch.	Zeitschrift für physiologische Chemie.
Z. Rüb.	Zeitschrift des Vereines für Rübenzuckerindustrie.

DETERMINATION OF RADICLES IN CARBON COMPOUNDS.

Chapter I.

INTRODUCTORY. DETERMINATION OF HYDROXYL (-OH).

THE quantitative analysis of inorganic compounds, as usually performed, consists almost exclusively in the determination of ions, since in the present state of the science this generally suffices for the identification of the substance; but to attain the same end in the case of organic bodies the elementary analysis requires supplementing by other methods. The percentage composition gives no information about the relative arrangement of the atoms in the molecule, but the demand for methods of analysis which will yield such knowledge increases with our growing insight into the constitution of carbon compounds. To supply this want certain "quantitative reactions" have been applied for the determination of special groups of atoms; they are widely, but almost exclusively, employed by technologists in the analysis of such substances as fats, waxes, resins, ethereal oils, caoutchouc, glue, paper, etc., and the results are known as the "acid number," "saponification number," "iodine number," "methoxyl number," "acetyl number," "carbonyl number," etc. The determination of such "numbers" or "values" obtained by the action of some reagent on a known weight of substance is frequently insufficient for scientific investigation, this renders it necessary to work out a special process for each group of organic compounds in order to determine the radicles which are present.

The reactions of organic compounds are only in part ionic; usually they are conditioned by the configuration and state of equilibrium of the molecule, and consequently a reaction which readily occurs with one compound may totally fail with another of very similar constitution on account of stereoisomerism; or, by substitution, a radicle may approximate more or less closely to the character and functions of another one. In these cases the quantitative separation of the compounds is more difficult, and can frequently only be accomplished by differences in crystallizing power, or by the preparation of derivatives which can be volatilized without decomposition.

Since the course of a particular reaction of an inorganic compound is only conditioned by the behavior of the ions which are to be determined, it follows that the analytical methods are in a sense independent of the nature of the compounds investigated, and consequently of very wide application. The matter is far otherwise with organic compounds: there are very few processes which, like Ziesel's method for determining methoxyl, can be applied almost universally.

Usually, then, it becomes necessary for the analyst himself to select the method most appropriate for his special purpose, or, perhaps by a combination of several, to devise one which may lead to the desired result. The successful methods hitherto proposed for the determination of organic radicles have been collected together in this work, and it is hoped that they may serve to indicate the direction in which research may be successfully prosecuted for the discovery of new ones applicable to hitherto unforeseen conditions.

DETERMINATION OF HYDROXYL (-OH).

The determination of the hydroxyl radicle in organic compounds consists in the preparation of derivatives by the following methods:

(I.) ACYLATION.—This consists in the introduction into the hydroxyl compound of the radicle of one of the acids mentioned below:

Acetic acid;
Benzoic acid and its substitution products;
Phenylsulphonic acid.

Of less frequent employment are the radicles of

Propionic acid;
Isobutyric acid;
Phenylacetic acid.

(II.) ALKYLATION.—Confined usually to the preparation of *benzyl* derivatives.

(III.) The preparation of CARBAMATES.

(IV.) The formation of ESTERS OF PHENYLCARBAMIC ACID.

As a rule, attention is first directed to the preparation of an acetyl or benzoyl derivative, the former usually by Liebermann & Hörmann's method (see page 7), the latter by that of Lossen or Schotten-Baumann (see page 18). Not infrequently, however, it becomes necessary to resort to one of the other forms of procedure in order to determine the constitution of the body under investigation. As the groups NH_1, NH_2 and SH are all capable of acylation, care is required to avoid confusion if the original compound contains nitrogen or sulphur. Instances are known of acetylation taking place in the absence of hydroxyl and of the groups just referred to, thus diacetylhydroquinol is formed from quinone, acetic anhydride, and sodium acetate;[1] tetrachloroquinone and acetyl chloride yield diacetyltetrachlorohydroquinol;[2] whilst pyrogallophthaleïn (galleïn), which contains only two hydroxyl groups, forms a tetracetyl and tetrabenzoyl derivative.[3] Acetylating reagents frequently cause isomerization or polymerization, and sometimes lead to the production of anhydrides, etc.; thus benzhydrylacetocarboxylican hydride is obtained from the isomeric orthocinnamocarboxylic acid by the action of acetic anhydride and sodium acetate,[4] and cantharic acid when heated in a sealed tube with acetyl chloride yields isocantharidin.[5] In view of these and similar facts, care should be taken to hydrolyse the presumptive acetyl derivative and identify the product with the original substance; should this

---

[1] Sarauw, B. 12, 680.   [3] Graebe, Ann. 146, 13.
[2] Buchka, B. 14, 1327.   [4] Benedikt and Ehrlich, M. 9, 529.
[5] Anderlini and Ghiro, B. 24, 1998.

not be possible, then proof must be obtained that the derivative does actually contain the acid radicle, the introduction of which has been attempted.

I. METHODS OF ACETYLATION.

(1) PREPARATION OF ACETYL DERIVATIVES.

The following reagents are employed for the prepation of acetyl derivatives from organic compounds containing hydroxyl groups:

(A) *Acetyl chloride.*
(B) *Acetic anhydride, sodium acetate.*
(C) *Glacial acetic acid.*
(D) *Chloracetyl chloride.*

(A) Acetylation by Means of Acetyl Chloride.

(*a*) Many hydroxyl derivatives react with acetyl chloride when simply mixed or digested on the waterbath. It is convenient to dissolve the substance and the chloride in benzene, and boil the solution until the evolution of hydrochloric acid ceases. If there is no danger of the hydrogen chloride causing secondary reactions (hydrolysis), of which an interesting case has been recorded,[1] the substance may be heated with the chloride in a sealed tube without solvent. Certain dibasic hydroxy acids of the aliphatic series, such as mucic acid, which are not changed with acetyl chloride alone, frequently react with it on the addition of zinc chloride.[2] In general, it may be stated that acetyl

---

[1] Herzig and Schiff, B. **30**, 397. Cf. Bamberger and Landsiedl, M. **18**, 307.
[2] S., p. 258.

chloride only reacts readily with alcohols and phenols, but, as it may lead to the production of anhydrides from polybasic acids, these are usually employed in the form of esters, which has the additional advantage of yielding products that are much more easily distilled than the corresponding derivatives of the acids themselves.[1]

(*b*) The following method[2] is frequently more convenient than the "acid" acetylation just described. The substance is dissolved in ether or benzene, and digested with the necessary quantity of acetyl chloride and dry alkali carbonate, the latter being in the proportion necessary to form a hydrogen salt as represented by the equation:

$$R.OH + CH_3.COCl + K_2CO_3 \longrightarrow R.O.CO.CH_3 + KCl + KHCO_3.$$

(*c*) Acetylation by means of acetyl chloride and aqueous alkali is described on p. 20.

(*d*) It is often convenient to allow the acetyl chloride to react with the compound under investigation in pyridine solution.[3]

(*e*) Diacetylacetone could only be acetylated by allowing its barium salt to react with acetyl chloride at the ordinary temperature.[4]

(*f*) Instead of acetyl chloride phosphorus trichloride, or preferably the oxychloride, or phosgene may be employed; they are allowed to react on a mixture of the substance and acetic acid in the proper propor-

---

[1] Wislicenus, Ann. **129**, 17.   [2] L. Claisen, B. **27**, 3182.
[3] A. Deninger, B. **28**, 1322.   [4] Feist, *Ibid.* **28**, 1824.

tion.[1] Thus, for example, phenol is readily acetylated by heating it at 80° with an equimolecular proportion of acetic acid and adding phosphorus oxychloride ($\frac{1}{3}$ molecule) gradually, by means of a dropping funnel. When hydrogen chloride is no longer evolved the product is poured into cold aqueous soda solution; after further washing with highly dilute alkali it is treated once with water, dried by means of calcium chloride, and distilled.

(B) Acetylation by Means of Acetic Anhydride.

(a) The substance is usually boiled with 5–10 parts of anhydride, or heated with it in a sealed tube during several hours.

(b) Not infrequently the substances must only be allowed to react during a short time, at a comparatively low temperature. Bebirine, for instance, is readily acetylated when digested with the anhydride during a short time at 40°–50°, but by its prolonged action amorphous substances are formed.[2]

(c) The substance may be mixed with an equal weight of dry sodium acetate, and 3–4 parts of the anhydride, and boiled for a short time in a reflux apparatus;[3] in the case of small quantities of substance 2–3 minutes boiling may suffice. The action appears to depend on the production of a sodium salt of the compound under examination, which then reacts with the anhydride. This method yields, on the whole, the most trustworthy results of any, and sel-

---

[1] J. pr. 25, 282; 26, 62; 31, 467.   [2] B. 29, 2057.
[3] C. Liebermann and O. Hörmann, *Ibid.* 11, 1619.

dom fails to give completely acetylated derivatives. It fails in the case of the α-hydroxyl of the hydroxyquinolines,[1] though these compounds yield benzoyl derivatives.

(*d*) A mixture of acetic anhydride and acetyl chloride may be used, or the action of the anhydride may be started by means of a drop of concentrated sulphuric acid.[2]

(*e*) The addition of zinc chloride[3] and of stannic chloride[4] has also been recommended.

(C) Acetylation by Means of Glacial Acetic Acid.

Acetylation, especially that of alcoholic hydroxyl groups, may often be accomplished by heating the substance with glacial acetic acid, under pressure if necessary; the addition of sodium acetate is also advantageous, and, in some cases, this is the only method which gives the desired result. Thus, camphorpinacone yields a chloride when treated with acetyl chloride, and is not changed by boiling acetic anhydride, but when it is boiled with glacial acetic acid for a short time, a stable acetyl derivative is formed, and an isomeric "labile" one by the action of the acid at the ordinary temperature during twenty-four hours.[5]

(D) Acetylation by Means of Chloracetyl Chloride.

This reagent has also been employed occasionally.[6]

---

[1] J. Diamant, M. **16**, 770. Cf. La Coste and Valeur, B. **20**, 1822.
[2] Franchimont, B. **12**, 1941.
[3] Franchimont, C. r. **89**, 711; B. **12**, 2058.
[4] H. A. Michael, B. **27**, 2686. [5] Beckmann, Ann. **292**, 17.
[6] Klobukowsky, B. **10**, 881. Cf. *Ibid.* **31**, 2790.

II. ISOLATION OF THE ACETYL DERIVATIVES.

Acetyl derivatives are isolated by pouring the product of the reaction into water. The excess of acetic acid may also be removed by the addition of methylic alcohol to convert it into methylic acetate, which is then volatilized; residual acetic anhydride is separated by distillation under reduced pressure. Acetyl derivatives, soluble in water, may often be precipitated by the addition of solid sodium carbonate, or by extracting the solution with chloroform or benzene. Ethylic acetate frequently proves to be an excellent medium for the subsequent recrystallization of the acetyl product.

III. DETERMINATION OF THE ACETYL GROUPS.

The various acetyl derivatives of a compound usually differ little in percentage composition, so that elementary analysis seldom affords information as to the number of acetyl groups which have entered the original molecule; thus, the mono-, di-, and tri-acetyl derivatives of the trihydroxybenzenes have an identical percentage composition. In such cases the acetyl groups must be eliminated and the acetic acid formed determined directly or indirectly.

(A) Hydrolytic Methods.

The following reagents are employed for the hydrolysis of acetyl compounds:

(a) *Water.*
(b) *Potassium hydroxide, sodium hydroxide.*
(c) *Barium hydroxide.*
(d) *Magnesia.*
(e) *Hydrochloric acid.*
(f) *Sulphuric acid.*
(g) *Hydriodic acid.*

(a) Some acetyl derivatives are hydrolysed by heating with water under pressure; thus butenyltriacetin, $C_4H_7(C_2H_3O_2)_3$, is completely hydrolysed by heating it with forty parts of water at 160° in a sealed tube, and the liberated acetic acid may be titrated.[1] Diacetylmorphine also loses one aceytl group by boiling it with water,[2] and acetyl dihydroxypyridine is still more unstable.[3]

(b) Hydrolysis by means of potassium hydroxide or sodium hydroxide is specially useful for the analysis of fats. The compound (1–2 grams) is gently boiled on the water-bath during fifteen minutes, in a wide-necked flask of 100–150 cc capacity, with alcoholic potash (25–50 cc) of known strength, which should be about N/2. During the heating the neck of the flask is covered with a cold funnel; at the conclusion of the hydrolysis phenolphthaleïn is added, and the excess of alkali determined by means of N/2 hydrochloric acid.[4] The method may also be employed for the determination of the molecular weight of the aliphatic alcohols. This is obtained

---

[1] Lieben and Zeisel, M. 1, 835.
[2] Wright-Becket, Journ. Ch. Soc. 12, 1033. Danckwortt, Arch. Pharm. 226, 57.
[3] M. 18, 619.   [4] Benedikt and Ulzer, M. 8, 41.

from the expression $M = \dfrac{56100}{V} - 42$, where M is the molecular weight, and V the number of milligrams of potassium hydroxide required to hydrolyse 1 gram of the acetyl derivative. If the compound is affected by air, the hydrolysis is carried out in an atmosphere of hydrogen;[1] should the original compound be insoluble in dilute hydrochloric acid, the acetyl derivative may be boiled with aqueous potash, the product acidified, and the precipitate weighed.[2]

(*c*) Barium hydroxide may be employed in many cases where potash causes decomposition, thus hæmatoxylin yields formic acid when boiled with highly dilute alkali, but barium hydroxide readily hydrolyses its acetyl derivatives without further decomposition.[3] One method of procedure[4] is to boil the compound under investigation with the hydroxide during 5–6 hours in a reflux apparatus. The product is filtered, the filtrate treated with carbonic anhydride in excess, again filtered, and the filtrate evaporated. The residue is dissolved in water, the liquid filtered, and, after washing, the barium in the filtrate is determined as sulphate. Since all the above operations are conducted in glass vessels, some alkali from these may neutralize a portion of the acetic acid and a correction thus becomes necessary. This is obtained by concentrating the filtrate from the barium sulphate in a platinum dish;

---

[1] Klobukowsky, B. **10**, 882.
[2] Vortmann, "Anleitung zur chemischen Analyse organischer Stoffe," p. 59.
[3] Erdmann and Schultz, Ann. **216**, 234.   [4] Herzig, M. **5**, 86.

when the excess of sulphuric acid has been volatilized, the residue is treated with pure ammonium carbonate until its weight becomes constant. It is now dissolved in water, the silica removed, and the sulphates in the filtrate determined as barium sulphate, the weight of which is added to that first found. If the hydrolysis, etc., can be carried out in vessels of silver,[1] the above correction is unnecessary. The action of the barium hydroxide solution is promoted by the previous addition to the substance of a few drops of alcohol.[2]

(*d*) Magnesia is generally employed in the following manner:[3] Ordinary "ignited magnesia," and the basic carbonate (magnesia alba) are both unsuitable, as they contain alkali carbonates which are difficult to remove. The magnesia is prepared from the sulphate or chloride, which must be free from iron; the solution is treated with alkali hydroxide in quantity insufficient to cause complete precipitation; after thorough washing the magnesia is retained as a paste under water. The acetyl derivative (1–1.5 grams) is intimately mixed with the magnesia paste (about 5 grams) and a little water, and transferred, together with water (100 cc), to a flask of resistant glass. The mixture is boiled in a reflux apparatus during 4–6 hours, although usually the hydrolysis is completed in 2–3 hours. The liquid is concentrated in the flask to a third of its original volume, cooled, filtered by means of a pump, the insoluble portion washed, and the filtrate and washings treated with ammonium

---

[1] Lieben and Zeisel, M. **4**, 42; **7**, 69.
[2] Barth and Goldschmiedt, B. **12**, 1237.
[3] H. Schiff, *Ibid.* **12**, 1531. Ann. **154**, 11.

chloride, ammonium hydrate, and ammoniacal sodium phosphate. The magnesium ammonium phosphate, after standing during twelve hours, is filtered, dissolved in dilute hydrochloric acid, and reprecipitated by means of ammonium hydrate; 1 part of $Mg_2P_2O_7 =$ 0.774648 parts of C_2H_2O. The solubility of magnesia in highly dilute solutions of magnesium acetate is too small to require a correction. Even "insoluble" acetyl derivatives may be hydrolysed by magnesia, provided that they are in a finely divided state, the boiling being prolonged to twelve hours if necessary. The magnesia method is advantageous in cases where the use of alkali causes decomposition and the production of colored substances which render titration uncertain.

(*e*) If hydrochloric acid (sulphuric acid) is without action on the hydroxyl compound, the acetyl derivative is heated with a known quantity of N/1 acid in a sealed tube or pressure-flask at 120°–150°, and the liberated acetic acid titrated.[1]

(*f*) Hydrolysis by means of sulphuric acid is especially advantageous when the original substance is insoluble in it. The acid employed should be free from oxides of nitrogen and contain 75 parts of concentrated acid in 32 parts of water. The dilute acid (10 cc) is mixed in a flask with a weighed quantity of the acetyl derivative (about 1 gram), which, if necessary, may be previously moistened with three or four drops of alcohol; the mixture is warmed on a hot but not boiling water-bath during a half hour, diluted with

---

[1] Schützenberger, Ann. de Ch. Ph. **84**, 74. Herzfeld, B. **13**, 266. Schmoeger, *Ibid.* **25**, 1453.

eight volumes of water, then boiled during 3-4 hours on the water-bath, and allowed to remain during twenty-four hours at the ordinary temperature. The precipitated hydroxyl derivative is then collected on a filter.[1,2] Should the hydroxyl derivative not be completely insoluble in the dilute acid a blank experiment must be made and the correction introduced.[3]

(*g*) Hydriodic acid has also been employed for the hydrolysis of acetyl derivatives.[3]

(B) Additive Method.[4]

This may be regarded as complementary to the method described under *f*. In cases where the acetyl derivative is insoluble in cold water, and the acetylation proceeds quantitatively, the yield of product from a given weight of hydroxyl compound gives a measure of the number of acetyl groups introduced. This method has recently been applied to the investigation of the acetylation products of tannic acid.[5]

(C) Weighing the Potassium Acetate.[6]

This is applicable to compounds yielding potassium salts insoluble in absolute alcohol. The acetyl derivative (1-2 grams) is boiled with a slight excess of potassium hydroxide solution until it is com-

---

[1] Liebermann, B. **17**, 1682. Herzig, M. **6**, 867-890.
[2] Ciamician and Silber, B. **28**, 1395.
[3] Ciamician, *Ibid.* **27**, 421, 1630.
[4] Goldschmiedt and Hemmelmayr, M. **15**, 321.
[5] H. Schiff, Ch. Ztg. **20**, 865.   [6] Wislicenus, Ann. **129**, 175.

pletely hydrolysed, water being added to replace that evaporated. The remaining alkali is neutralized with carbonic anhydride, the liquid evaporated as completely as possible on the water-bath, and the residue thoroughly extracted with absolute alcohol. The alcoholic solution is evaporated to dryness and the residue again extracted, any insoluble matter being removed and well washed, and the liquid evaporated in a tared vessel. The dried potassium acetate remaining is then cautiously fused, allowed to cool over sulphuric acid, and weighed.

(D) Distillation Method.

Fresenius[1] first suggested that the acetic acid from acetates could be liberated with phosphoric acid and determined by distillation, with or without the help of steam. The method was then applied by various chemists to the hydrolysis of acetyl derivatives, but since they replaced the phosphoric acid by sulphuric acid their results were not satisfactory.[2] Subsequently the use of phosphoric acid was again proposed.[3] The acetyl product is hydrolysed by means of alkalis or barium hydroxide, acidified at the ordinary temperature with phosphoric acid, filtered, and well washed; the filtrate and washings are then distilled until the distillate is completely free from acid, fresh water being introduced into the retort from

---
[1] Z. anal. Ch. 5, 315; 14, 172.
[2] Erdmann and Schulze, Ann. 216, 232. Buchka and Erk, 18, 1142. Schall, *Ibid.* 22, 1561.
[3] Herzig, M. 5, 90.

time to time as may be necessary. The distillation is at first carried out over a flame and subsequently from an oil-bath, the temperature being allowed to rise to 140°–150°, or a water-bath may be employed, in which case the pressure is reduced.[1] The connections must all be of caoutchouc, as corks would absorb acetic acid, and the alkali and acid employed must be free from nitrates or nitrites. The presence of chlorides is not hurtful, as these do not liberate hydrogen chloride in presence of the phosphoric acid, which is one advantage it possesses over sulphuric acid.[2] The distillate is treated with baryta water in excess, and concentrated in a platinum dish, the excess of barium removed by means of carbonic anhydride, and the filtrate evaporated to dryness; water is then added, the liquid filtered, the insoluble portion well washed, and the barium in the filtrate and washings determined as sulphate, 1 gram $BaSO_4 = 0.5064$ gram $C_2H_3O_2$ or 0.5070 gram $C_2H_4O_2$.

The acetyl groups in acetylated gallic acids[3] were determined by mixing the substance (3–4 grams) with pure alcohol (5 cc) and sodium hydroxide (2–3 grams) dissolved in water (15 cc). After the hydrolysis was completed, the alcohol was dissipated, the residue acidified with phosphoric acid, the acetic acid driven over in a current of steam, and its amount determined by titrating the distillate with sodium hydrate solution, phenolphthaleïn being used as indicator. One source of error in this method arises from carbonic

---

[1] H. A. Michael, B. 27, 2686.
[2] R. and H. Meyer, *Ibid.* 28, 2967.
[3] P. Sisley, Bull. Soc. Chim. III. 11, 562. Z. anal. Ch. 34, 466.

anhydride, which is always present in the sodium hydrate, and is often produced by the hydrolysis itself; it naturally volatilizes together with the acetic acid. The difficulty may be avoided by heating the neutralized liquid to boiling, adding a very small quantity of N/1 acid, again boiling, and then neutralizing, the process being repeated until the neutralized liquid ceases to become red on boiling; this shows that all the carbonates are decomposed and no loss of acetic acid need be apprehended. It has been suggested[1] that, after the hydrolysis, elimination of the alcohol, and acidification by means of phosphoric acid, the liquid should be boiled in a reflux apparatus until the carbonic anhydride is removed, the subsequent operations being similar to those above described. Sources of error in this method are described on p. 26.[2]

BENZOYL DERIVATIVES.

(1) PREPARATION OF BENZOYL DERIVATIVES.

The following reagents are employed for the introduction of the benzoyl radicle into hydroxyl compounds:

Benzoyl chloride;
Benzoic anhydride, sodium benzoate;
p-Brombenzoyl chloride, p-Brombenzoic anhydride;
o-Brombenzoyl chloride;
m-Nitrobenzoyl chloride;
Phenylsulphonic chloride.

---

[1] P. Dobriner, Z. anal. Ch. **34**, 466, foot-note.
[2] Cf. Goldschmiedt and Hemmelmayr, M. **14**, 214; **15**, 319.

(A) Preparation of Benzoyl Derivatives by Means of Benzoyl Chloride.

(*a*) The "acid" method consists in heating the substance with the chloride at 180° during several hours in a reflux apparatus; it is not advisable to employ a sealed tube unless there is assurance that the hydrochloric acid will not cause secondary reactions nor, in the case of nitrogenous compounds, combine with them to form hydrochlorides which would then cease to react;[1] when this may occur the calculated quantity of chloride is employed, and the heating continued during about four hours at 100°–110°.

(*b*) The preceding method has been largely superseded by the use of the chloride in dilute aqueous alkaline solution.[2] It has been widely applied,[3] is usually known as the Schotten-Baumann method, and seldom fails to give good results. The substance is well shaken with sodium hydroxide solution (10%) and benzoyl chloride in excess until the smell of the latter is no longer noticeable.[4] If the benzoylation is to be as complete as possible more concentrated alkali should be used, say fifty parts of soda (20%) and six parts of the chloride in a closed flask.[5] The temperature should not exceed 25°,[6] and it is frequently desirable to add the alkali and chloride alternately little by little, whilst in some cases the former must be highly

---

[1] Danckwortt, Arch. Pharm. **228**, 581.
[2] Lossen, Ann. **161**, 348; **175**, 274, 319; **205**, 282; **217**, 16; **265**, 148, foot-note.
[3] Baumann, B. **19**, 3218. [4] Baumann.
[5] Panormow, B. **24**, R. 971. [6] v. Pechmann, *Ibid.* **25**, 1045.

dilute.[1] It has also been found to be advisable to use the reagents in the proportion of seven molecules of soda and five of the chloride to each hydroxyl;[2] the alkali is dissolved in water (8–10 parts), and the shaking and gentle cooling continued during 10–15 minutes. For experiments with pyragallol the flask must be filled with coal-gas; in the case of similar substances which are so unstable in presence of caustic alkali, sodium carbonate,[3] bicarbonate, or sodium acetate may be used.[4] The precipitated benzoyl derivatives are usually white and semi-solid, and gradually harden and crystallize by prolonged contact with water; often traces of benzoyl chloride or benzoic acid are retained with great tenacity. For the purification of the benzoyl derivative of dextrose[5] it was necessary to dissolve out the crude product with ether; this was distilled off, and the residue treated with alcohol, which decomposed the last portions of benzoyl chloride that had not been removed by prolonged shaking of the ethereal solution with concentrated alkali. The alcoholic liquid was treated with soda in excess, precipitated with water, and the alcohol and ethylic benzoate removed by means of steam. The residue was then repeatedly recrystallized; at first from alcohol, then from glacial acetic acid. The pure compound is insoluble in ether, whilst the crude preparation readily dissolves. Benzoic acid may be frequently removed by sublimation in vacuo, or by extraction with boiling

---

[1] B. **31**, 1598.
[2] Skraup, M. **10**, 390.
[3] Lossen, Ann. **265**, 148.
[4] Bamberger, M. & J., II., p. 546.
[5] Skraup, M. **10**, 395.

carbon bisulphide.¹ Repeated extraction with alkali is usually effective for the purification of benzoyl derivatives soluble in ether, but it may produce partial hydrolysis. Commercial benzoyl chloride often contains chlorobenzoyl chloride,² and since the chlorobenzoyl derivatives are less soluble than the benzoyl derivatives themselves, recrystallization is not adequate to secure a product free from chlorine. It appears also that pure benzoyl chloride may yield chloro-derivatives.³ Benzotrichloride may contain benzal chloride; during the conversion of the former into benzoyl chloride by the action of lead oxide or zinc oxide the latter may yield benzaldehyde, the presence of which would cause complications.⁴ Lactones often yield benzoyl derivatives of acids which are soluble in alkali; they are separated by acidifying and removing the benzoic acid from the precipitate by steam distillation.⁵

Schotten-Baumann's method has also been applied to the preparation of acetyl derivatives, but with comparatively little success on account of the greater instability of acetyl chloride.⁶

(*c*) Benzoyl derivatives may also be prepared in ethereal or benzene solution, with the help of dry alkali carbonate,⁶ or of tertiary bases such as quinoline, pyridine, or dimethyl aniline.⁷ (Cf. p. 6.)

(*d*) Sodium ethoxide⁸ may also be employed for the decomposition of benzoyl chloride, and it was only in

---

¹ Barth and Schreder, M. **3**, 800.
² V. Meyer, B. **24**, 4251.   Goldschmiedt, M. **13**, 55, foot-note.
³ B. **29**, 2057.   ⁴ Hoffmann and V. Meyer, *Ibid.* **25**, 209.
⁵ *Ibid.* **30**, 127.   ⁶ *Ibid.* **27**, 3183.
⁷ L. Claisen, *Ibid.* **31**, 1023.   ⁸ L. Claisen.

this manner that the benzoyl derivative of diacetyl-acetone could be obtained.[1] The ketone was heated in a reflux apparatus during six hours, with two molecular proportions each of benzoyl chloride and sodium ethoxide, which had been dried at 200°; after cooling, the sodium chloride and benzene were removed, the residue dissolved in ether, and the solution shaken with dilute alkali.

(*e*) Pyridine may be used in place of aqueous, or alcoholic alkali.[2] The product is triturated with dilute hydrochloric acid and recrystallized from alcohol.

(B) Preparation of Benzoyl Derivatives from Benzoic Anhydride.

(*a*) The hydroxyl compound is heated with benzoic anhydride, in an open vessel, at 150° during 1–2 hours.[3]

(*b*) In some cases the use of benzoic anhydride and sodium benzoate produces a more complete acylation than Schotten-Baumann's method.[4] As an example of its use, scoparin (2 grams), benzoic anhydride (10 grams), and dry sodium benzoate (1 gram) were heated in an oil-bath at 190° during six hours; the product was treated at the ordinary temperature overnight with aqueous sodium hydroxide (2%), and the precipitated hexabenzoyl derivative purified by means of alcohol.

---

[1] Feist, B. 28, 1824. [2] Deninger, *Ibid*. 28, 1322.
[3] Liebermann, Ann. 169, 237.
[4] Goldschmiedt and Hemmelmayr, M. 15, 327.

Not infrequently the addition of sodium benzoate is quite unnecessary.[1]

(C) Preparation of Substituted Benzoic Acid Derivatives and of Phenylsulphonic Chloride.

(*a*) *Parabromobenzoyl chloride.*[2]— Parabromobenzoic acid is intimately mixed with the equivalent quantity of phosphorus pentachloride, and warmed until the evolution of hydrogen chloride slackens. The product is then fractionated under reduced pressure; the pure compound melts at 42°, boils at 174° (102 mm), and is readily soluble in benzene and light petroleum.

(*b*) *Parabromobenzoic anhydride*[3] is prepared by heating sodium parabromobenzoate (3 parts) with parabromobenzoyl chloride (2 parts) at 200° during an hour. It melts at 212°, is almost insoluble in ether, carbon bisulphide, and glacial acetic acid, dissolves slightly in benzene, and is purified by recrystallization from chloroform.

(*c*) *Orthobromobenzoyl chloride*[4] is prepared in a similar manner to its isomer. It is a liquid, boiling at 241°–243°, and may be distilled under the ordinary pressure without decomposition.

(*d*) *Metanitrobenzoyl chloride*[5] is formed from the nitrobenzoic acid by gradually and intimately mixing with it the requisite amount of phosphorus penta-

---

[1] Arch. Pharm. **235**, 313.   [2] B. **21**, 2244.
[3] Schotten and Schlömann, *Ibid.* **24**, 3689.
[4] B. **21**, 2244.   Schöpf, *Ibid.* **23**, 3436.
[5] Claisen and Thompson, *Ibid.* **12**, 1943.

chloride; the phosphorus oxychloride is removed by distillation, and the residue fractionated under reduced pressure. It melts at 34° and boils at 183°–184° (50–55 mm).

(e) *Phenylsulphonic chloride*[1] is obtained by heating sodium phenylsulphonate with phosphorus pentachloride in equivalent proportions; when the action ceases the product is poured into water, the oily portion removed, washed with water, dissolved in ether, and the solution decolorized by treatment with animal charcoal. The compound melts at 14° and boils at 120° (10 mm).

(D) Acylation by Means of Substituted Benzoic Acid Derivatives and of Phenylsulphonic Chlorides.

(a) *Parabromobenzoyl chloride, or parabromobenzoic anyhdride,* has been used for acylation, the number of the original hydroxyl groups, being determined from the bromine content of the product.[2]

(b) *Orthobromobenzoyl chloride*[3] and *metanitrobenzoyl chloride*[4] are also well adapted for the determination of hydroxyl groups.

(c) *Phenylsulphonic chloride*[5] has been employed for

---

[1] Otto, Z. f. Ch. *1866*, 106.

[2] F. Loring Jackson and G. W. Rolfe, Am. Chem. Journ. **9**, 82; B. **20**; R. 524.

[3] Schotten, *Ibid.* **21**, 2250.

[4] Claisen and Thompson, *Ibid.* **12**, 1943. Schotten, *Ibid.* **21**, 2244.

[5] Hinsberg, B. **23**, 2962. Schotten and Schlömann, *Ibid.* **24**, 3689.

the same purpose; it is either allowed to act like the benzoyl chloride in the Schotten-Baumann method, or it is warmed with the hydroxyl compound (phenol) and zinc dust, or zinc chloride.[1]

Phenylsulphonic derivatives are often more stable than the corresponding benzoyl compounds.[2]

(2) ANALYSIS OF BENZOYL DERIVATIVES.

(*a*) The exact number of benzoyl groups in many benzoyl derivatives is shown by their elementary analysis; in substitution products the amount of haloid, nitrogen, or sulphur is determined.

(*b*) The following method has been suggested for the direct determination of the benzoic acid:[3] The substance (about 0.5 gram) is hydrolysed by heating it during two hours at 100°, in a sealed tube, with concentrated hydrochloric acid (10 parts), which has been saturated with benzoic acid at the ordinary temperature. The product is allowed to remain 1–2 days at the ordinary temperature, filtered by means of the pump, and the precipitate washed, at first with more of the hydrochloric acid, then with a saturated aqueous solution of benzoic acid. The purified benzoic acid is now dissolved in N/10 sodium hydroxide solution in excess, titrated with excess of acid, and the neutralization effected with the needful quantity of the soda solution. The latter is standardized against pure benzoic acid,

---

[1] C. Schiaparelli, Gazz. 11, 65.   [2] B. 30, 669.
[3] G. Pum, M. 12, 438.

phenolphthaleïn being employed as the indicator. The admixture of the acid and water during the washing of the benzoic acid always causes a precipitation of benzoic acid, so that the results obtained by this method are invariably about 1 per cent. too high; therefore, this amount must be deducted from the percentage of acid found, or the exact correction ascertained by means of a blank experiment with the same quantities of liquids as have been used in the main one.

(c) A method of more general application consists in separating the benzoic acid from the hydrolysed substance by means of a current of steam, and titrating the distillate;[1] its principle is therefore identical with the determination of acetyl groups, and it presupposes that the compound under examination is completely hydrolysed by alkalis, and yields no acid other than benzoic, volatile with steam. The substance (about 0.5 gram) is mixed with alcohol (30–50 cc) and potassium hydroxide in excess, and heated in a reflex apparatus; when the hydroylsis is completed the product is cooled, acidified with concentrated phosphoric acid solution or vitreous phosphoric acid, and distilled in a current of steam. The distillation is conducted slowly at first, and alcohol added, if necessary, by means of a dropping funnel, the object being to secure the gradual deposition, in a crystalline state, of the hydrolysis products as otherwise resinous substances might surround the benzoic acid and considerably

---

[1] R. and H. Meyer, B. 28, 2965.

hinder its volatilization. When the distillate measures 1–1.5 liters, the following 150 cc are collected separately and tested for benzoic acid by titration, and, as soon as it is no longer present, the distillation is stopped. The combined distillate is rendered alkaline with a known quantity of N/10 sodium hydrate solution, standardized against pure benzoic acid, and evaporated in a platinum, silver, or nickel dish to a volume of 100–150 cc, when the excess of alkali is titrated back, the liquid being boiled to expel carbonic anhydride; this may be regarded as accomplished when boiling during ten minutes produces no change in the indicator, which is aurin or rosolic acid. In order to guard against the production of sulphites and sulphates, the concentration of the alkaline liquid is carried out by means of a spirit or petroleum lamp, unless a special gas burner is available.

(*d*) Benzoylmorphine has been examined by direct titration.[1] The substance was dissolved in methylic alcohol, mixed with a little water, normal sodium hydrate solution (100 cc) added, and boiled in a reflux apparatus until a portion of it gave no turbidity with water; titration with normal hydrochloric acid, in presence of phenolphthaleïn, showed that the original compound was the monobenzoyl derivative. The same method was successfully applied to the analysis of dibenzoylpseudomorphine and tribenzoylmeythlpseudomorphine.

---

[1] Vongerichten, Ann. **294**, 215. Cf. Knorr, B. **30**, 917–920.

ACYLATION BY MEANS OF OTHER ACID RADICLES.

Propionic anhydride, isobutyric anhydride, opianic acid,[1] stearic anhydride,[2] and phenylacetyl chloride are sometimes used for acylation, as their relatively high boiling points facilitate their reaction with the hydroxyl compound.

(*a*) *Propionyl* derivatives are prepared by heating the substance with propionic anhydride, in a stout, closed bottle, at 100° during two hours; an open vessel may also be employed, and the reaction started by the addition of a drop of concentrated sulphuric acid.[3]

(*b*) *Isobutyryl* derivatives are prepared in a similar manner. Isobutyryl ostruthin was prepared by heating ostruthin (3 grams) with isobutyric anhydride (10 grams) in a sealed tube at 150° during 3 hours. The product was poured into water, allowed to remain until it became crystalline, washed with warm water until neutral, pressed, dried by means of filter paper, and recrystallized from alcohol.[4]

(*c*) *Phenylacetyl chloride* is not difficult to prepare,[5] and is used[6] like benzoyl chloride in Schotten-Baumann's method, the substance being dissolved in dilute aqueous potassium hydroxide solution, and well shaken with excess of the chloride.

(*d*) The extent to which *phosphoric acid* may prove useful remains at present undetermined.[7]

---

[1] B. 31, 358.   [2] Ann. 262, 5.   [3] Arch. Pharm. 228, 127.
[4] Jassoy, Arch. Pharm. 228, 551.   [5] B. 20, 1389; 29, 1986.
[6] Hinsberg, *Ibid.* 23, 2962.   [7] *Ibid.* 30, 2368; 31, 1094.

II. ALKYLATION OF HYDROXYL GROUPS.

The hydroxyl of phenol and primary alcohols is capable of alkylation, and the number of alkyl groups introduced may be determined from the resulting ethers by Zeisel's method (cf. p. 33). As a rule, the phenolic ethers are not hydrolysed by alkalies (cf. p. 457), hence it is possible to differentiate between the hydroxyl and carboxyl of the hydroxy acids. It has, however, been shown that the use of potassium hydroxide and alkyl iodides may lead to the production of compounds with the alkyl directly linked to carbon,[1] and that, on the other hand, hydroxyl in the ortho-position relative to carbonyl oxygen is determinable by acylation, but not by alkylation.[3]

Diazomethane may also be used as an alkylating agent.[4]

PREPARATION OF BENZYL DERIVATIVES.

Benzyl ethers of phenols are prepared by heating the latter in a reflux apparatus, during several hours, with the calculated quantities of sodium ethoxide and benzyl chloride in alcoholic solution, the precipitated sodium chloride is removed by filtration from the hot liquid,[5] and the composition of the

---

[1] B. **30**, 2368; **31**, 1094.

[2] Herzig and Zeisel, M. **9**, 217, 882; **10**, 144, 735; **11**, 291, 311, 413; **14**, 376.

[3] Graebe. Herzig, M. **5**, 72. Schunk and Marchlewsky, Journ. Chem. Soc. **65**, 185. Kostanecki, B. **26**, 71, 2901. Perkin, Journ. Chem. Soc. **67**, 995; **69**, 801.

[4] v. Pechmann, B. **28**, 856; **31**, 64, 501, Ch. Ztg. **98**, 142.

[5] Haller and Guyot, C. r. **116**, 43.

ether determined by elementary analysis. Benzyl iodide may also be employed for the preparation of these compounds.[1]

III. PREPARATION OF CARBAMATES BY MEANS OF CARBAMYL CHLORIDE.

PREPARATION OF CARBAMYL CHLORIDE.[2]

Ammonium chloride is placed in a distillation flask attached to a long and wide condenser, heated at about 400° in an air bath, and treated with a current of carbonyl chloride, dried by means of sulphuric acid. The carbamyl chloride, which has a highly offensive smell, distils over and condenses to a colorless liquid, or to long, broad needles melting at 50°. It volatilizes at 61°–62°, and, after prolonged standing, polymerizes to cyamelide, for which reason it should be employed as quickly as possible after its preparation. In contact with water or moist air, it is hydrolysed to carbonic anhydride and ammonium chloride.

PREPARATION OF CARBAMATES.

Carbamyl chloride reacts with hydroxyl derivatives in accordance with the equation:

$$NH_2CO.Cl + HO.R \longrightarrow NH_2.CO.OR + HCl,$$

the resulting carbamates readily crystallize.[3] It is

---

[1] M. & J. II., p. 125.
[2] Gattermann & G. Schmidt, B. **20**, 858.
[3] Gattermann, Ann. **244**, 38.

usually only necessary to mix the substances in equivalent proportion in ethereal solution as the reaction generally proceeds quantitatively at the ordinary temperature; in the case of some polybasic phenols gentle warming is requisite. The amount of nitrogen in the product is a measure of the number of hydoxyl groups in the original compound. Great excess of the chloride should not be used, as it may lead to the production of ethereal allophanates, $NH_2.CO.NH.CO.OR$.

IV. PREPARATION OF DIPHENYLCARBAMYL CHLORIDE $(C_6H_5)_2N.CO Cl$.

This substance has been found especially useful in the investigation of rhodinol (geraniol).[1] It is prepared by dissolving diphenylamine (250 grams) in chloroform (700 cc), adding anhydrous pyridine (120 cc), and passing a current of carbonyl chloride (147 grams) into the liquid, which is maintained at 0°. After remaining during 5–6 hours, the chloroform is distilled off on the water-bath, and the residue crystallized from alcohol (1.5 liters). The yield is 300 grams, the product, after recrystallization from alcohol (1 liter), is pure, and melts at 84°.[2]

---

[1] Erdmann and Huth, J. pr. **53**, 45.
[2] *Ibid.* **56**, 7

V. PREPARATION OF PHENYLCARBAMIC ACID DERIVATIVES.

PREPARATION OF PHENYLISOCYANATE.[1]

Commercial phenylurethane (15 grams) is mixed with phosphoric anhydride (30 grams) in a small retort, and heated by means of a luminous flame, a large distillation flask being employed as receiver; the combined distillate from a number of such preparations is then fractionated once. The isocyanate boils at 169° (769 mm),[2] and the yield is 52–53 per cent.[3]

ACTION OF PHENYLISOCYANATE ON HYDROXYL DERIVATIVES.[4]

Ethereal phenylcarbamates are formed by the interaction of hydroxyl compounds and phenylisocyanate in equimolecular proportion in accordance with the equation:

$$R.HO + C_6H_5N:CO \rightarrow C_6H_5NH.CO.OR.$$

The reaction often proceeds at the ordinary temperature, but it is best to rapidly boil the compounds, mixed in the requisite proportion, by means of a previously heated sand-bath, and complete the reaction by shaking and gentle warmth.[5] Polybasic phenols are

---

[1] H. Goldschmiedt, B. **25**, 2578, foot-note.
[2] Hofmann, B. **18**, 764. [3] Zanoli, *Ibid.* **25**, 2578, foot-note.
[4] Hofmann, Ann. **74**, 3; B. **18**, 518. Snape, *Ibid.* **18**, 2428.
[5] Tessmer, *Ibid.* **18**, 969.

heated in a sealed tube during 10–16 hours;[1] if the compound eliminates water at this temperature the phenylisocyanate is converted by it into carbonic anhydride and carbanilide.[2] The duration of the boiling in an open vessel should be shortened as much as possible to reduce the production of diphenylcarbamide. When cold the product of the reaction is treated with a little benzene or ether to dissolve unaltered phenylisocyanate, then, after the removal of the benzene or ether, washed with cold water and recrystallized from alcohol, ethylic acetate, or a mixture of ether and light petroleum, which leaves the sparingly soluble diphenylcarbamide undissolved. The presence of negative groups in the molecule of the hydroxyl derivative hinders or completely prevents the reaction; thus trinitrophenol gives no derivative when heated at 180° under pressure.[3]

An attempt has been made to determine the presence of hydroxyl groups by the use of 1:2:4-*chlordinitrobenzene.*[4]

---

[1] Snape, B. 18, 2428.
[2] Tessmer, *Ibid.* 18, 969. Beckmann, Ann. 292, 16.
[3] Gumpert, J. pr. 31, 119; 32, 278.
[4] Vongerichten, Ann. 294, 215.

CHAPTER II.

DETERMINATION OF METHOXYL, $\overset{\mathrm{I}}{\mathrm{CH_3O}}$-, ETHOXYL, $\overset{\mathrm{I}}{\mathrm{C_2H_5O}}$-, AND CARBOXYL, $\overset{\mathrm{I}}{\mathrm{CO.OH}}$.

I. DETERMINATION OF METHOXYL, $\overset{\mathrm{I}}{\mathrm{CH_3O}}$-.

S. ZEISEL'S METHOD.[1]

This method, which is distinguished for beauty and reliability, depends on the conversion of the methyl of the methoxy group into methyl iodide by means of hydriodic acid, the methyl iodide being then decomposed by alcoholic silver nitrate solution into silver iodide. The original apparatus, represented in Fig. 1, consists of a reversed condenser K, through which water at 40°–50° flows; at the lower end a flask A of 30–35 cc capacity is attached by means of a cork; the flask has a side tube sealed on through which a current of carbonic anhydride may be passed. A Geissler's potash bulb is connected to the upper end of the condenser, also by means of a cork; it contains 0.25–0.5 gram red phosphorus suspended in water, and is maintained at a temperature of 50°–60° by the water-bath in which it is placed. Its object is to absorb any iodine or hydriodic acid which might be carried over by the

---

[1] M. **6**, 989; **7**, 406.

methyl iodide vapor. The two flasks which complete the apparatus have a capacity of 80 cc each, the first

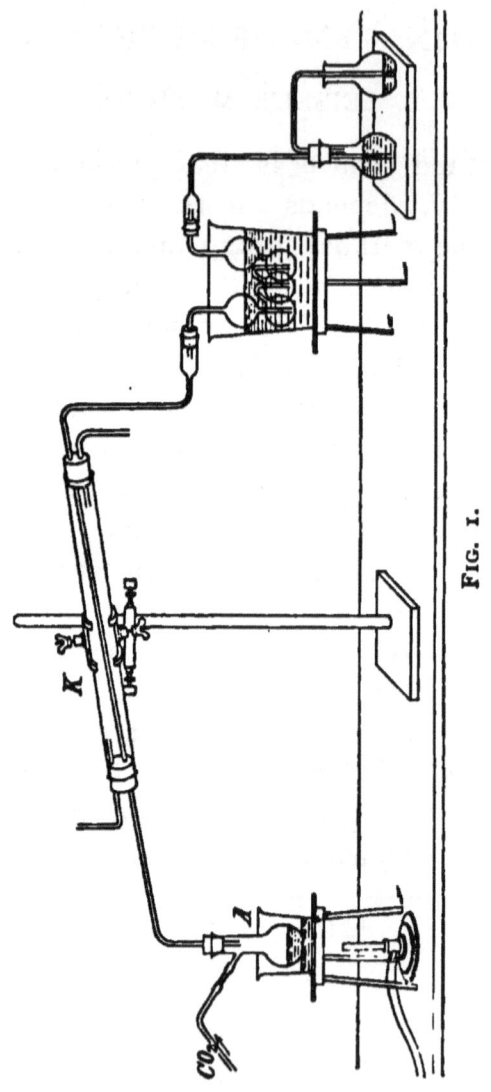

FIG. 1.

contains 50 cc of alcoholic silver nitrate solution, the second 25 cc; they are connected by means of corks, and may be conveniently replaced by two distillation flasks, the side tube of the first being bent downwards at a right angle into the second. A modified apparatus has been described, which serves as a combined condenser and washing arrangement,[1] and also a second one, which has in addition a very convenient appliance for heating and supplying the water to the condenser.[2] Modified boiling flasks,[3] (Fig. 2), which prevent the action of the heated hydriodic acid on the cork, have been designed. If the substance under examination is not volatile, the condenser may be replaced by a vertical tube bent back in a U shape. The method is not applicable to compounds containing sulphur, and the hydriodic acid employed must not have been prepared by means of hydrogen sulphide, otherwise it is difficult to completely free it from volatile sulphur compounds, the presence of which would be apt to cause the formation of mercaptanes and silver sulphide. C. A. F. Kahlbaum, of Berlin, supplies "hydriodic acid for methoxyl determination," which is prepared by means of phosphorus, and is trustworthy. Should a blank experiment show that the hydriodic acid produces a perceptible precipitate in the silver nitrate solution, it must be purified by distillation, the first and last quarters of the distillate being rejected; it should have a sp. gr. = 1.68–1.72. Boiling the acid with a reversed condenser, even

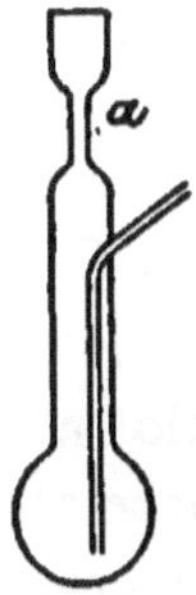

FIG. 2.

---

[1] Benedikt and Grüssner, Ch. Ztg. 13, 872.
[2] L. Ehmann, *Ibid.* 14, 1767; 15, 221.
[3] Benedikt, *Ibid.* 13, 872. M. Bamberger, M. 15, 505.

during several days,[1] does not suffice for its purification. The *silver nitrate* solution is prepared by dissolving the fused salt (2 parts) in water (5 parts) and adding absolute alcohol (45 parts); it is kept in the dark, and the quantity required for each determination filtered into the absorption flasks.

I. METHOD FOR NON-VOLATILE SUBSTANCES.

After the apparatus is put together, tested, and found to be air-tight, the silver nitrate solution is introduced into the absorption flasks, and the substance (0.2–0.3 gram), together with the hydriodic acid (10 cc), placed in the distillation flask; unless Bamberger's pattern is employed this should also contain a few pieces of porous plate to regulate the ebullition; it is then heated to boiling in a glycerine-bath. During this time the current of carbonic anhydride is passed through the apparatus at the rate of three bubbles in two seconds. The gas employed must be washed with water, and also with silver nitrate solution, to remove any hydrogen sulphide arising from impurities in the marble. The warm water must also be supplied to the condenser and the bath containing the potash bulbs. Some 10–15 minutes after the acid begins to boil the silver nitrate becomes turbid and soon a white double compound of silver nitrate and silver iodide precipitates in the first flask; the liquid in the second one usually remains clear, but sometimes becomes opalescent if the current of carbonic anhydide is very rapid or the substance

---

[1] Benedikt.

particularly rich in methoxyl groups, these conditions may also cause the precipitate to become yellow. The conclusion of the experiment is readily indicated by the complete subsidence of the precipitate, which becomes crystalline; the time required is 1–2 hours. The tubes and flasks with the silver solution are disconnected, and the second one diluted with five parts of water; if no precipitate appears after remaining several minutes nothing more is done to it, otherwise it is added to the contents of the first flask, which are poured into a beaker, any precipitate adhering to the tubes is removed to the beaker by means of a feather and jet of water; the volume is now made up to about 500 cc with water, evaporated to one half on the water-bath, then water and a drop of nitric acid added, and the liquid digested until the silver iodide is completely precipitated; it is then filtered and weighed in the usual manner. The precipitate adhering to the tubes is usually dark-colored, possibly from the presence of a trace of phosphorus, but this does not affect the accuracy of the determination. 100 parts of silver iodide $= 13.20$ parts of $CH_4O = 6.38$ parts of CH_4. The method is applicable to compounds containing chlorine, bromine,[1] or nitro-groups, but not to sulphur compounds.[2] In the case of nitro-derivatives, or other compounds which readily liberate iodine from hydriodic acid, it is desirable to place a little red phosphorus in the boiling-flask. The potash bulbs require refilling after four or five determinations. Hydriodic acid causes many sub-

---

[1] G. Pum, M. 14, 498.
[2] Zeisel. *Ibid.* 7, 409. Benedikt and Bamberger, *Ibid.* 12, 1.

stances to become resinous, and the resin may protect a portion of the methoxy compound from the action of the acid. This difficulty may be overcome by adding to the acid acetic anhydride (6–8 volumes per cent), as was shown in the case of methyl and acetylethylquercetin, rhamnetin, and triethoxyphloroglucinol.[1] The method is also well adapted for the determination of alcohol of crystallization.[2]

2. MODIFICATIONS OF THE METHOD FOR ITS USE WITH VOLATILE COMPOUNDS.

Volatile substances may usually be treated in the manner described above if, at the commencement of the experiment, a slow stream of carbonic anhydride is employed and cold water run through the condenser. The following special modifications for particularly volatile compounds have been suggested.[3] The substance (0.1–0.3 gram) is sealed into a small bulb of thin glass, which is sealed up in a larger tube together with hydriodic acid (10 cc, sp. gr. $= 1.7$) and a piece of heavy glass about 2 cm in length with a sharp corner. The heavy glass is to assist in breaking the bulb with the substance before the heating, but is unnecessary if the latter is enclosed in testtube glass with a long capillary. The larger tube is 30–35 cm long and 1.2–1.5 cm inner diameter; both ends are drawn out, the one to fit into a tube 10 cm long and 1–2 cm inner diameter, which is

---

[1] Herzig, M. **9**, 544. Cf. Pomeranz, *Ibid.* **12**, 383.
[2] J. Herzig and H. Meyer, *Ibid.* **17**, 437.
[3] Zeisel, *Ibid.* **7**, 406.

sealed to a wider tube, the other so that a piece of stout rubber tube will fit over it quite tightly. The drawn-out ends must be strong enough to resist the pressure during the heating, and sufficiently thin to be readily broken after being scratched with a file. The substance and hydriodic acid are heated at 130° during two hours, then, when cold, one point of the tube is fitted into the narrow one mentioned above, the wide portion of which passes through a triply bored cork into a wide-mouthed flask. Into the second opening of the cork the condenser fits, whilst the third contains a piece of stout glass rod bent to a Z form; by turning this the drawn-out end of the heating-tube is broken. The contents are transferred to the flask partly by shaking, partly by gently warming; the upper capillary is covered with a piece of rubber, the end broken, and a current of carbonic anhydride immediately passed through the apparatus. The determination then proceeds in the manner already described.

3. MODIFICATION OF ZEISEL'S METHOD.[1]

Instead of red phosphorus and water the potash bulbs contain a solution consisting of arsenious anhydride (1 part), potassium carbonate (1 part), and water (10 parts). The bulbs must be refilled for each determination to prevent the apparatus becoming choked with precipitated anhydride, but this is compensated for by the fact that not the slightest reduction (black-

---

[1] J. Gregor, M. 19. 116.

ening) of the silver nitrate solution takes place. The N/10 *silver nitrate solution* is made by dissolving the nitrate (17 grams) in water (30 cc) and diluting to a liter with commercial absolute alcohol, its titer being determined by means of N/10 potassium thiocyanate solution. For the alkyloxyl determination the silver solution (75 cc) is acidified with a few drops of nitric acid, free from nitrous acid, and divided between the two absorption flasks. At the conclusion of the experiment the silver solution with the precipitate is diluted with water to 250 cc, cautiously shaken, filtered by means of a dry ribbed filter into a dry flask, and 50 cc or 100 cc of the clear filtrate acidified with nitric acid, free from nitrous acid, treated with ferric sulphate solution, and titrated in the ordinary manner.[1]

METHOD FOR THE DIFFERENTIATION OF METHOXYL AND ETHOXYL.

Zeisel's method does not distinguish between methoxyl and ethoxyl; should this be necessary, the alkyl iodide must be prepared in quantity sufficient for its identification, or, if possible, Lieben's iodoform test must be applied. For the differentiation of the alkyls the investigation of the action of phenyl isocyanate on the alkyloxy derivatives has been suggested.[2] The compound is heated with phenyl isocyanate, in equimolecular proportion, at 150° during several hours in a sealed tube. The product is steam-distilled and the volatile portion purified by recrystal-

---

[1] Volhard, J. pr. **9**, 217. Ann. **190**, 1. Z. anal. **13**, 171; **17**, 482.
[2] Beckmann, Ann. **292**, 9, 13.

lization from a mixture of ether and light petroleum; methylphenylurethane melts at 47°, ethylphenylurethane at 50°, and they can be further distinguished by analysis.

DETERMINATION OF ETHOXYL ($C_2H_5\overset{\shortmid}{O}-$).

The determination of ethoxyl[1] is carried out exactly in the manner described in the preceding section for methoxyl except that the water in the condenser and in the bath surrounding the potash bulbs should be heated at about 80°. 100 parts of silver iodide = 19.21 parts C_2H_5O = 12.34 parts C_2H_5.

DETERMINATION OF CARBOXYL ($C\overset{\shortmid}{H}.OH$).

The following methods are employed for the determination of the basicity of organic acids:

(A) Analysis of metallic salts of the acid.
(B) Titration.
(C) Etherification.
(D) Determination of the electrolytic conductivity of the sodium salts.
(E) Indirect methods:
    (1) Carbonate method.
    (2) Ammonia method.
    (3) Hydrogen sulphide method.
    (4) Iodine method.

It is easy to decide which of these methods is the most suitable for any special case, but the qualitative differentiation between carboxyl and phenolic hy-

---

[1] Zeisel, M. **7**, 406.

droxyl frequently presents difficulties that can only be overcome with certainty by the preparation of the amide and its conversion into the nitrile.

(A) Determination of Carboxyl by Analysis of Metallic Salts of the Acid.

In many cases the number of carboxyl groups in an organic compound may be determined by the analysis of its neutral salts; of these the *silver salts* are usually the most appropriate, as they are generally formed directly without admixture of hydrogen salts and are almost always anhydrous. Exceptions to this rule are, however, encountered; thus the silver salts of cantharidinic acid,[1] camphoglycuronic acid,[2] and metaquinaldinic acid[3] crystallize with one, three and four molecules of water respectively, and hydrogen silver salts,[4] though not of frequent occurrence, are known. Aromatic hydroxymonocarboxylic acids containing two nitro-groups often give salts containing two atoms of silver. As examples may be mentioned 3:5-dinitrohydrocumaric, 1:3:5-dinitroparahydroxybenzoic and 2:6-dinitro-5-hydroxy-3-4-dimethylbenzoic acids.[5] Many silver salts are very sensitive to light or air, and some, like silver oxalate, are explosive; for the analysis of such the compound is dissolved or suspended in water or acid, and treated with hydrogen

---

[1] Homolka, B. **19**, 1083.
[2] Schmiedeberg and Meyer, Z. physiol. Chem. **3**, 433.
[3] Eckhardt, B. **22**, 276.
[4] A list of them is given in Lassar-Cohn "Manual of Organic Chemistry," translated by Alex. Smith, p. 345.
[5] W. H. Perkin, Jun., Journ. Chem. Soc. (*1899*), **75**, 176.

sulphide or hydrochloric acid. Silver salts which do not explode when heated are usually analyzed by ignition in a porcelain crucible; if the residual silver contains carbon it is dissolved in nitric acid, the solution diluted and filtered, and the silver precipitated by means of hydrochloric acid.

Pyridine and quinoline derivatives and amino-acids in particular often give characteristic *copper and nickel salts*, whilst, in the aliphatic series, the *zinc salts* may often be usefully employed. *Sodium, potassium, calcium, barium, magnesium*, and, less frequently, *leaa salts* are also sometimes used for the determination of basicity, but, as many acids do not yield well-defined neutral salts, and groups other than carboxyl can exchange hydrogen for metal, the method has not a very wide application.

(B) Titration of Acids.

The basicity of a carboxyl derivative may often be determined by titration if the molecular weight of the compound is known; N/10 sodium hydroxide, potassium hydroxide, or barium hydroxide may be used for the titration in aqueous solution, or, in the case of the first two, in alcoholic solution. N/2 ammonium hydroxide has also been employed.[1] The acids used are generally hydrochloric or sulphuric, but the latter is unsuited for work with alcoholic solutions, as the precipitation of insoluble sulphates prevents a correct observation of the end reaction. The liquid, alcohol, ether, etc., in which the compound under examination

---

[1] Haitinger and Lieben, M. 6, 292.

is dissolved must either be free from acids or must be previously accurately neutralized by means of N/10 alkali. Phenolphthaleïn, methyl orange, rosolic acid, curcumin, or litmus, are usually employed as indicators, the first two more frequently than the others. If the liquid is dark colored the use of "alkali blue" is often convenient, and attention must always be paid to the possible presence of carbonic anhydride. A somewhat curious and interesting attempt has been made to determine the neutrality by taste.[1]

(C) Etherification.

In very many cases carboxylic and phenolic hydrogen may be differentiated by the etherification of the compound with alcohol and hydrogen chloride. It has, however, been shown[2] that acids with the group

$$\underset{\underset{t}{C}\ \underset{t}{C}}{\overset{t'}{C}.COOH}$$

(t and t′ = tertiary carbon atom) do not yield esters with alcohol and hydrogen chloride if both the carbon atoms marked t are linked to Cl, Br, I, or NO_2, whilst the groups of smaller mass F, CH_3, OH in the same positions greatly retard, but do not entirely prevent, etherification. On the other hand, certain phenols such as phloroglucinol,[3] which gives a diether, hydroxyanthracene (anthrol and α- and β-naphthol[4]) yield ethers when treated with hydrogen

---

[1] T. W. Richards, Am. Chem. Journ., **20**, 125.

[2] V. Meyer and others. Many papers appeared on the subject beginning B. **27**, 510, and ending **29**, 2569.

[3] *Ibid.* **17**, 2106; **21**, 603.

[4] Liebermann and Hagen, *Ibid.* **15**, 1427.

chloride and alcohol. The etherification is most conveniently carried out by boiling the substance during 3–5 hours in a reflux apparatus with a large excess of absolute alcohol containing 3–5 per cent of hydrogen chloride or sulphuric acid.[1] Occasionally alcohol of 95 per cent may be employed if more sulphuric acid is used.[2] Some substances form additive compounds with alcohol and hydrogen chloride.[3] This, as also the contamination of the ester by traces of chlorine derivatives, which can only be removed with difficulty, may lead to confusion.

The esters obtained by acid or alkaline etherification are, in general, distinguished from the phenolic ethers by the ease with which aqueous or alcoholic alkalis hydrolyse them, but exceptions are known since trinitromethoxybenzene (methoxy picrate) when boiled with concentrated potassium hydroxide yields methyl alcohol and potassium picrate,[4] and methoxyanthracene (methyl anthranol) is also decomposed by boiling with alcoholic potash.[5] The composition of the esters is determined by elementary analysis, and the alkyloxy groups by the methods described in the earlier portion of this chapter.

---

[1] E. Fischer and A. Speier, B. **28**, 3252.
[2] Bishop Tingle and A. Tingle, Am. Chem. Journ. **21**, 243.
[3] Freund, B. **32**, 171.
[4] Ann. **174**, 259.
[5] Liebermann and Hagen, B. **15**, 1427.

(D). Determination of the Basicity of Acids by means of the Electrolytic Conductivity of the Sodium Salts.

It has been shown that the degree of electrolytic conductivity of the sodium salt is a certain indication of the basicity of the corresponding acid.[1] The method is of very general application, since insoluble acids usually yield sodium salts which dissolve in water, but it fails in the case of acids which are so feeble that their sodium salts are hydrolysed by water sufficiently to impart an alkaline reaction to the solution. The following apparatus is required for the determination:

(1) A small *induction coil* (J, Fig. 5), such as is employed for medicinal purposes, and which requires only one or two cells for prolonged use. The spring of the interrupter must vibrate rapidly so as to produce a high pitched sound in the telephone, as this is more easily heard than a deeper tone.

(2) A *bridge* consisting of a scale 100 cm in length divided into millimeters; along it stretches a wire provided with a sliding contact. The wire is of platinum, German silver, platinoid, or manganin, of which the last is the best on account of its low temperature coefficient. The wire must be calibrated.[2]

(3) A *rheostat* for adjusting the resistance (W, Fig. 5).

---

[1] Ostwald, Z. **2**, 901; **1**, 74. Valden, *Ibid.* **1**, 529; **2**, 49.

[2] Strouhal and Barus, Wied. Ann. **10**, 326. The method is also described by Jones, "Freezing-point, Boiling-point, and Conductivity Methods," Chem. Pub. Co., 1897.

(4) A *resistance cell* for the electrolyte (E, Fig. 5). Kohlrausch's form (Fig. 3) is used for low resistances whilst that of Arrhenius (Fig. 4) is employed for dilute solution where the resistance is high. The electrodes must be platinised by filling the vessel with a dilute solution of hydroplatinochloric acid and passing a current of 4–5 volts.

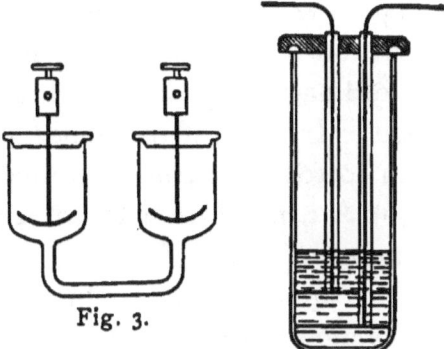

Fig. 3.

Fig. 4.

The direction of the current is changed occasionally and the electrolysis continued until both electrodes are completely covered with platinum black, which only requires a short time; the platinum chloride in the cell is now replaced by sodium hydroxide solution, the electrolysis continued for a few moments, the electrodes then thoroughly and carefully washed with hydrochloric acid, and finally with water; the sodium hydroxide removes all chlorine which is otherwise very obstinately retained by the platinum. The use of Lummer and Kurlbaum's solution for platinizing is highly recommended, as the tone minima are much more distinct.[1] The solution consists of platinum chloride (1 part), lead acetate (0.008 part), and water (30 parts); it is electrolysed with a current density of 0.03 ampères per sq. cm., the direction of the current being frequently changed and continued until each

---

[1] Kohlrausch, Wied. Ann. 1897, p. 315; E. Cohen, Z. **25**, 1611.

electrode has been the cathode during at least fifteen minutes.

(5) A *telephone*. Ostwald states that the most sensitive ones are made by Ericsson of Stockholm, but for ordinary purposes a Bell instrument is sufficiently good. In using it the unoccupied ear may be closed with cotton to exclude external sounds.

(6) A *water-bath* with *stirrer* and *thermometer*, or a *thermostat*.[1]

The apparatus is arranged in the form of Kirchhoff's modification of the Wheatstone bridge (Fig. 5), the connections being made with stout copper wire. The induction coil is enclosed in a sound tight case, or is placed in another room. If determinations of solutions of a substance at different concentrations are to be made the solution is most conveniently prepared in the resistance cell itself, portions are then withdrawn by means of an accurately calibrated pipette, and the desired volume of water added which has previously been brought to the necessary temperature in the thermostat. As a rule the telephone does not give an absolutely sharp minimum at any given point, but it is easy to find two limits beyond either of which the tone rises; these are usually separated by an interval of 0.5–2 mm, and the required position is taken as midway

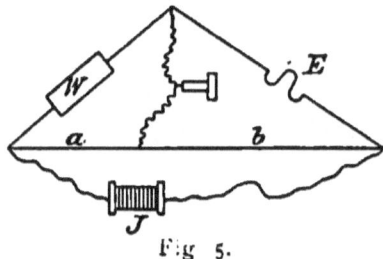
Fig 5.

---

[1] Ostwald, Z. **2**, 564, where also a good description of the other parts of the apparatus is given.

between them. A little experience enables the conductivity to be determined with an accuracy of 0.1 per cent. If the tone minimum becomes indistinct the electrodes must be replatinised. The conductivity is calculated from the measurements by means of the formula $\mu = k \cdot \dfrac{v \cdot a}{w \cdot b}$, where

μ = the molecular conductivity;
v = the volume of the solution in liters which contains a gram molecule of the electrolyte;
w = the adjusting resistance;
a = the length of wire to the left of the sliding contact (Fig. 5);
b = that to the right of the contact (Fig. 5)
k = the resistance of the cell.

The value of k is determined by measuring the conductivity of N/50 solution of potassium chloride, for which Kohlrausch found the values:

$$\mu = 112.2 \text{ at } 18°.$$
$$\mu = 129.7 \text{ at } 25°.$$

Other solutions may also be used.[1] The value $\dfrac{b}{a}$ for a wire 1000 mm in length has been calculated by Obach and an abbreviated table of the results is given in the appendix. The conductivity of the water employed, which should be as highly purified from dissolved substances as possible, is determined in the same manner as that of the solution, the value for

---

[1] Wiedemann and Ebert, Physik. Praktikum, p. 389.

each liter (v) is calculated according to the formula, and subtracted from the uncorrected value of μ. For basicity determinations the conductivity is usually determined at concentrations of one gram molecule in 32 and 1024 liters respectively. The mean difference Δ between these values is as follows:

Monobasic acids.... $\Delta = 10.4 = 1 \times 10.4$
Dibasic   "   .... $\Delta = 19.0 = 2 \times 9.5$
Tribasic   "   .... $\Delta = 30.2 = 3 \times 10.1$
Tetrabasic   "   .... $\Delta = 41.1 = 4 \times 10.3$
Pentabasic   "   .... $\Delta = 50.1 = 5 \times 10$

A method has been described[1] for determining the basicity of acids based on the alterations which they exhibit in electrolytic conductivity on the addition of alkali.

Instead of the telephone and induction coil, a double commutator and a galvanometer may be used to determine the electrolytic conductivity, the commutating apparatus, termed a secohmmeter, is so arranged that one commutator is included in the battery circuit and the other in that of the galvanometer; on rotating the current is reversed in the liquid so frequently that polarization is annulled whilst the galvanometer is commuted.[2]

---

[1] D. Berthelot, C. r. **112**, 287.
[2] Cahart and Patterson, "Electrical Measurements," p. 109.

(E) Indirect Methods for the Determination of the Basicity of Acids.

These methods may be divided into four classes according to the nature of the substance liberated by the acid:

(1) *Carbonate method.*
(2) *Ammonia method.*
(3) *Hydrogen sulphide method.*
(4) *Iodine-oxygen method.*

(1) *Carbonate Method.*—The substance (0.5–1 gram) is dissolved in water in a flask closed by a rubber stopper with three holes. In one hole a condensing tube is fitted, which, at the lower end, is flush with the stopper whilst the upper end is connected with an absorption apparatus consisting of two calcium chloride tubes and potash bulbs. Through the second hole a tube passes to the bottom of the flask, the end being drawn out and bent upwards; by means of this tube a current of air, free from carbonic anhydride, is passed. The third hole of the flask is closed with a small dropping funnel, the end of which is also drawn out and bent upwards and dips below the liquid in the flask. The solution of the acid is gently boiled, and barium carbonate, in the form of a thin paste, is added in small quantities by means of the funnel. When the operation is completed the apparatus is allowed to cool in a current of purified air, again boiled, cooled, and the absorption bulbs weighed.[1] A similar method,

---

[1] Goldschmiedt and Hemmelmayr, M. 14, 210.

based on the decomposition of sodium hydrogen carbonate, has also been described.[1]

(2) *Ammonia Method.* The acid (about 1 gram) is dissolved in excess of alcoholic potassium hydroxide, and made up to 250 cc with alcohol of the same strength (93 per cent). The excess of alkali is neutralized by carbonic anhydride, the precipitated carbonate and bicarbonate filtered off and washed with 50 cc of alcohol (98 per cent). The alcohol is removed from the filtrate and washings by distillation, and the residue boiled with 100 cc of ammonium chloride solution (10 per cent). The potassium salt of the acid decomposes the ammonium chloride, and the liberated ammonia is determined in the usual manner. The amount of alkali carbonate dissolved by 100 cc of alcohol (93 per cent) is equivalent to 0.34 cc of normal acid; a correction for this must be applied and also one for the ammonium chloride hydrolised by the water; this is determined by a blank experiment, 100 cc of the solution being boiled during the same length of time, 1–2 hours, as in the actual determination.[2]

The method gives good results with the feebler fatty acids, and is especially useful when the dark color of the solution prevents direct titration.

(3) *Hydrogen Sulphide Method.*[3] Compounds containing carboxyl liberate hydrogen sulphide from certain metallo-hydrogen sulphides when allowed to react in

---

[1] Vohl B. 10, 1807. C. Jehn, *Ibid.* 10, 2108.
[2] P. C. McIlhiney, J. Am. 16, 408.
[3] F. Fuchs, M. 9, 1132, 1143; 11, 363.

an atmosphere of hydrogen sulphide, according to the equation:

$$NaSH + R.COOH + xH_2S \longrightarrow RCOONa + H_2S + xH_2S$$

two volumes of hydrogen sulphide being liberated for each volume of hydrogen, replaceable by metal, in the original compound. Hydroxyl hydrogen in phenols, alcohols, and hydroxy-acids does not react with the metallo-hydrogen sulphides.

Preparation of the Solution.

The majority of alkali salts are sparingly soluble in solutions of the hydrosulphides, hence the solution of the latter must not be so concentrated as to hinder the reaction from being rapidly completed. Potassium hydroxide solution, not exceeding 10 per cent, is boiled with baryta water in excess, the flask closed, and the liquid allowed to cool and deposit barium carbonate. The clear solution is now poured into the vessel to be used for the analysis, and saturated with hydrogen sulphide.

Method of Analysis.

The evolved hydrogen sulphide may be determined:

(a) *volumetrically;*
(b) *by titration.*

The former method is the easier and is therefore generally employed.

(a) *Volumetric Determination.*

This method is based on the same principle as Victor Meyer's vapor density determination. The apparatus, Fig. 6, consists of a long-necked flask A, made of thick glass; it is fitted with a rubber stopper c through which the delivery tube B passes, this is wide at one end but terminates in a capillary at the other. The second hole of the stopper is closed by means of a glass rod from which the vessel containing the substance is suspended. Previous to the determination the greater portion of the flask is filled with hydrogen sulphide, but the upper portion of the neck and the delivery tube contain air which is expelled by the evolved hydrogen sulphide and collected over water in a graduated tube. The substance under examination is dried, finely powdered, and about 0.5 gram weighed into the small vessel, the glass rod being pressed into the rubber stopper as far as the mark 1, the vessel fitted on to it by means of the stopper, which, with the delivery tube, is passed air-tight into the flask. The apparatus is allowed to remain for a few moments to equalize the

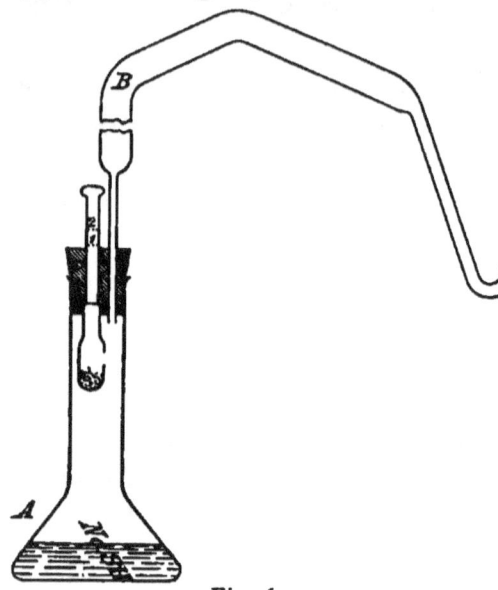

Fig. 6.

temperature, then the capillary end of the delivery-tube is dipped into water below the open end of the gas-measuring vessel, and the vessel with the substance dropped into the sulphide solution by pushing in the rod to the mark 2, care being taken not to alter the position of the stopper itself. The evolution of hydrogen sulphide ceases after a few minutes. The same solution may be employed for a second or third determination, but each time the delivery-tube must be previously filled with dry air. The weight of carboxyl hydrogen G is calculated from the results by the formula:

$$G = \frac{\frac{1}{2}V(b-W)}{760(1 + 0.00366t)} \cdot 0.0000896$$
$$= \frac{V.(b-w).0.00000005895}{1 + 0.00366t},$$

where $V =$ the observed volume of air displaced in cc, $b =$ the height of the barometer, and w the tension of aqueous vapor at the observed temperature t.

(b) *Titration Method.*

The apparatus employed consists of a short-necked flask A, Fig. 7, fitted with a rubber stopper and glass rod exactly as used in the preceding method, but the delivery-tube is short in order to expedite the expulsion of air. Before the stopper, with the substance adjusted in the manner described above, is inserted into the flask, tartaric

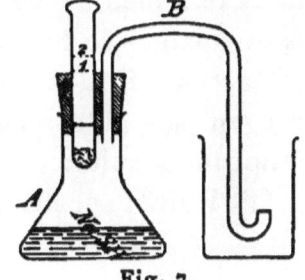

Fig. 7.

acid or oxalic acid (about 0.25 gram) is dropped into the potassium hydrogen sulphide solution and the stopper immediately inserted air-tight as shown. As soon as the solution of gas ceases the beaker represented in the figure is replaced by a smaller one containing concentrated potassium hydroxide solution. Some of this rises in the tube on account of the absorption of the gas, but the error so introduced compensates itself at the end of the experiment. As soon as the beaker of alkali has been put into position, the substance is dropped into the sulphide solution with the same precautions as observed in the preceding method; after the cession of the gas evolution, which contimes during 1–5 minutes, the pressure is adjusted by lowering the beaker, the contents are poured into a large flask, and the beaker and evolution tube washed. The alkali and washings are diluted to about 500 cc, neutralized with acetic acid, and titrated with iodine solution in presence of starch. Since

$$H = H_2S = I_2,$$

the iodine required, divided by 2 × 126.5, gives the weight of the replaceable hydrogen. The error due to the insertion of the glass rod from mark 1 to 2 may be determined by means of a blank experiment, but it is so small as to be usually negligible. More recently the action of substituted phenols, etc., on alkali hydrogen sulphides has been investigated[1] with the following results:

(1) Haloïd substituted phenols with one hydroxyl

---

[1] Fuchs, M. 11, 363.

group are without action on the sulphides, but if *two* hydroxyl groups are present *one* reacts.

(2) Only the *paramononitro*-phenols react.

(3) Under certain conditions the presence of carboxyl groups causes the phenolic hydroxyl to decompose the sulphides.

(4) In general lactones do not react, but lactone-acids may suffer partial resolution.[1]

With the above exceptions the method provides a ready means of differentiating carboxylic hydrogen from phenolic or alcoholic, a distinction which the two preceding methods do not furnish with certainty.

(4) *Iodine-oxygen Method.*[2] This depends on the fact that even feeble organic acids liberate iodine from potassium iodide and potassium iodate in accordance with the equation:

$$6R.COOH + 5KI + KIO_3 \rightarrow 6R.COOK + 3I_2 + 3H_2O.$$

The liberate diodine, in presence of alkali, evolves oxygen from hydrogen peroxide:

$$I_2 + 2KOH \rightarrow KOI + KI + H_2O \text{ and}$$
$$KOI + H_2O_2 \rightarrow KI + H_2O + O_2.$$

The oxygen may be measured in a modified Wagner & Knop's Azotometer,[3] or in any other convenient vessel.

The *apparatus* consists of an evolution flask, with a small cylinder of about 20 cc capacity fused to the middle of the bottom inside, and a large glass cylinder

---

[1] H. Meyer, M. **19**, 715.
[2] Baumann and Kux, Z. Anal. Ch. **32**, 129.   [3] *Ibid.* **13**, 389.

with two communicating burettes and a thermometer fastened to the interior of the cover. The cylinder and burettes are filled with water, the latter by connecting them with a flask from which water is forced by air pressure from a hand blower, the connecting tube being provided, if needful, with a stop-cock. The evolution flask is closed by means of a rubber stopper carrying a tube with a stop-cock which is connected with the graduated burette below the stop-cock in which the latter terminates, and which is used for adjusting the pressure. The temperature of the evolution flask is equalized before and after the determination by placing it in water of the same temperature as that in the large cylinder enclosing the burettes. The following *reagents* are required:

(1) Potassium iodide ⎱
(2)   "   iodate ⎰ Absolutely free from acid.
(3) Hydrogen peroxide 2–3 per cent. solution.
(4) Aqueous potassium hydroxide solution (1:1).
(5) Distilled water, recently boiled and free from carbonic anhydride.

The *determination* is carried out in the following manner: The acid (0.1–0.2 gram) is mixed with finely divided potassium iodate (about 0.2 gram), potassium iodide (2 grams), and water (40 cc) in a bottle provided with a well-fitting stopper, and allowed to remain at the ordinary temperature during twelve hours, or at 70°–80° during a half hour, until the iodine is completely precipitated. The solution is now transferred to the outer portion of the evolution flask, the bottle being washed with not more than 10 cc water. Into the inner cylinder of the evolution flask is poured by means of a funnel a mixture consisting of hydrogen peroxide (2 cc) and potash solution (4 cc), made immediately before use and cooled to the ordinary temperature. The evolution flask is now closed with its stopper and allowed to stand in water during ten minutes, the stopcock of the burette being opened to equalize the pressure; at the end of this time it is closed, the level adjusted to the zero mark, and if after five minutes no change takes place the experiment is proceeded with, otherwise the cooling is continued during another five minutes. When equilibrium is established 30–40 cc of water are run from the burette in order to reduce the pressure, the evolution flask is removed from the cooling vessel by means of a cloth and rotated so that the liquids at first circulate without mixing and are then suddenly brought into contact. The shaking is continued vigorously for a short time and the flask then returned to the cooling vessel. The evolution of oxygen begins at once, and is completed in a few seconds; after about ten minutes the pressure in the two burettes is adjusted and the volume read the number of cc of gas, multiplied by the value in the table [1] in the appendix corresponding to the pressure and temperature, gives directly the weight of carboxylic hydrogen. An *iodometric method* for the determination of acids has also been described, for details the original paper should be consulted.[2]

---

[1] Baumann, Z. f. ang. Ch. *1891*, p. 328
[2] M. Gröger, *Ibid. 1890*, pp. 353, 385.

Chapter III.

DETERMINATION OF CARBONYL ($\overset{\text{II}}{\text{CO}}$).

The presence of the carbonyl group in aldehydes ketones, etc., is recognized by the preparation of derivatives of the following compounds:

(1) Phenylhydrazine.
(2) Hydroxylamine.
(3) Semicarbazide.
(4) Amidoguanidine.
(5) Paramidodimethylaniline.

(I) CARBONYL DETERMINATION BY MEANS OF PHENYLHYDRAZINE.

The method is divisible as follows:

(A) *Preparation of phenylhydrazones from phenylhydrazine.*
(B) *Preparation of substituted hydrazones.*
(C) *Indirect method.*

(A) Preparation of Phenylhydrazones.[1]

Carbonyl compounds combine with phenylhydrazine forming water and phenylhydrazones,

$$C_6H_5NH.N:CRR_1;$$

dihydrazones, with the hydrazine groups linked to neighboring carbon atoms, are termed osazones. The reaction usually takes place most readily in dilute acetic

---

[1] E. Fischer, B. **16**, 661, 2241, foot-note; **17**, 572; **22**, 90.

acid solution, often at the ordinary temperature, almost always by heating on the water-bath. Frequently it is advisable to allow the reaction to proceed at the ordinary temperature in presence of concentrated acetic acid, which acts as a dehydrating agent and in which the phenylhydrazones, as a class, are sparingly soluble.[1] E. Fischer dissolves or suspends the substance in water or alcohol, and adds, in excess, a mixture of phenylhydrazine hydrochloride (1 part) and crystallized sodium acetate (1.5 parts) dissolved in water (8–10 parts). Free mineral acids must be neutralized by means of sodium hydroxide or sodium carbonate, as their presence hinders the reaction; the presence of nitrous acid is particularly hurtful and it must be removed by means of carbamide, as otherwise it reacts with the phenylhydrazine and forms diazobenzene imide and other oily products. Confusion may also be caused by the production of acetylphenylhydrazine from the dilute acetic acid.[2]

The phenylhydrazones gradually separate from the solution of their components in an oily or crystalline form, and, in the latter case, are purified by recrystallization from water, alcohol, or benzene.

It is often desirable to heat the compound under investigation with free phenyhydrazine, and increased pressure may be used if there is no danger of phenylhydrazides being formed.[3] The product is poured into water, the phenylhydrazone removed by filtration, washed with dilute hydrochloric acid to free it from excess of phenylhydrazine, and recrystallized; in some

---

[1] Overton, B. **26**, 20.   [2] Anderlini, *Ibid.* **24**, 1993, foot-note.
[3] M. **14**, 395.

cases glycerol is employed for washing, the last portions being removed by water.[1] Aliphatic ketones react readily in ethereal solution, and the water which is produced may be absorbed by recently ignited potassium carbonate or calcium chloride. In the case of ketophenols or ketoalcohols the hydroxyl group should be acetylated before treatment with phenylhydrazine; acids are usually used in the form of esters, but the sodium salt is sometimes employed[2] and the condensation promoted by the addition of a mineral acid.[3] Hydrazones may also be prepared from oximes.[4] The carbonyl group in many lactones and acid anhydrides condenses with phenylhydrazine,[5] but not with hydroxylamine;[6] on the other hand, many quinones, such as anthraquinone, do not react with phenylhydrazine or only with one molecular proportion, as in the case of naphthoquinone and phenanthraquinone, whilst some, such as benzoquinone and toluquinone, oxidize it to benzene.[7] Ortho-disubstituted ketones frequently do not react with phenylhydrazine,[8] and certain unsaturated ketoalcohols, such as ethylic acetoacetate[9] and ethylic camphoroxalate,[10] yield monophenylhydrazides, the ketonic group being un-

---

[1] Thoms, B. **29**, 2988.   [2] Bamberger, *Ibid.* **19**, 1430.
[3] Elbers, Ann. **227**, 353.
[4] Just, B. **19**, 1205. von Pechmann, *Ibid.* **20**, 2543, foot-note.
[5] R. Meyer and E. Saul, *Ibid.* **26**, 1271. Hemmelmayr, M. **13**, 667. Ephraim, B. **26**, 1376.
[6] v. Meyer and Münchmeyer, *Ibid.* **19**, 1706.   Hölle, J. pr. **33**, 99.   [7] S. p. 538.
[8] Baum, B. **28**, 3209. V. Meyer, *Ibid.* **29**, 830, 836.
[9] Nef. Ann. **266**, 52.
[10] Bishop Tingle, Am. Chem. Journ. **20**, 339. A. Tingle and Bishop Tingle, *Ibid.* **21**, 258.

affected. Hydroxyketones and aldehydes of the aliphatic series yield phenylosazones, a portion of the phenylhydrazine being simultaneously reduced to aniline and ammonia.[3]

A method has been described for the purification of commercial phenylhydrazine.[4]

(B) Preparation of Substituted Hydrazones.

The chief substitution product of phenylhydrazine which has hitherto been employed for the preparation of phenylhydrazones is the parabromo-derivative.

Preparation of Parabromophenylhydrazine.[5] Phenylhydrazine (20 grams) is poured into hydrochloric acid (200 grams, sp. gr. = 1.19) and the precipitated salt uniformly distributed throughout the liquid, which is cooled to 0°; bromine (22.5 grams) is now dropped in, the addition occupying 10–15 minutes, the liquid being well shaken during this time. After remaining during twenty-four hours the precipitate is removed, washed with a little cold hydrochloric acid, dissolved in water, and treated with sodium hydroxide in excess. The base separates in flocculent crystals which are extracted with ether, the ether evaporated, and the residue recrystallized from water. The hydrochloric acid mother liquor contains bromodiazobenzene chloride, which is reduced by the addition of stannous chloride (60 grams); the precipitate is separated, washed with concentrated hydrochloric acid, and treated with water and alkali, the base being collected and purified

---

[3] E. Fischer and Tafel, B. **20**, 3386. [4] B. Overton, *Ibid.* **26**, 19.
[5] Michaelis, *Ibid.* **26**, 2190.

in the manner described above. The yield is 80 per cent. Bromophenylhydrazine requires to be protected from light and air; it should be kept in the dark in well-stoppered colored bottles from which the air has been displaced by carbonic anhydride or coal-gas. In these circumstances, if the compound has been highly purified and dried, it may be retained for years without change; colored specimens may be readily purified by recrystallization from water, to which a few drops of sodium hydroxide should be added. The pure compound melts at 107°–109°, the acetyl derivative at 170°.[1]

Substituted Phenylhydrazones.

Parabromophenylhydrazine is well adapted for the identification of certain sugars, such as arabinose, and has also been used in the investigation of ionone and irone;[3] it is generally employed in acetic acid solution, care being taken to prevent the liquid from boiling, as, in these circumstances, acetyl parabromophenylhydrazine is formed.[4]

Paranitrophenylhydrazine also gives well defined condensation products with many aldehydes and ketones which serve for their identification. The reaction usually proceeds in aqueous solution with the hydrochloride, but the free base in alcohol or acetic acid may be employed.[5]

In addition the following substituted phenylhydrazines have been used for the production of

---
[1] Tiemann and Krüger.   [2] E. Fischer, B. 24, 4221, foot-note.
[3] Tiemann and Krüger, *Ibid.* 28, 1755.   [4] *Ibid.* 26, 2190.
[5] E. Bamberger & Kraus, *Ibid.* 29, 1834. Bamberger, *Ibid.* 32, 1806. E. Hyde, *Ibid.* 32, 1810.

DETERMINATION OF CARBONYL. 65

phenylhydrazones: *dibromo-*, *symmetrical tribromo-*, *tetrabromo-*, *parachloro-*, *pariodo-*, and *metadiiodo-*,[1] whilst some derivatives of *diphenylhydrazine* have also been described.[2]

(C) Indirect Method.[3]

This method depends on allowing the aldehyde or ketone to react with excess of phenylhydrazine; the excess, together with any hydrazide, is then oxidized by means of boiling Fehling's solution, the liberated nitrogen being collected; phenylhydrazones are not decomposed by this treatment.

The *reagents* required are as follows:

Copper sulphate solution (70 grams $CuSO_4.5H_2O$ in 1 liter). *Alkaline solution of sodium potassium tartrate*, made by dissolving 350 grams of the tartrate, and 260 grams potassium hydroxide in 1 liter; these two solutions are mixed in equal volumes to form the Fehling's solution.

Sodium acetate (10 per cent solution).

Phenyldrazine hydrochloride (5 per cent solution.)
The *analysis* is made by mixing the compound under examination (0.1–0.5 gram) with an accurately measured quantity of the phenylhydrazine hydrochloride solution (1 part) and the sodium acetate solution (1½ parts) in a 100 cc measuring flask. The phenylhydrazine hydrochloride is taken, if possible, in quantity sufficient to yield 15–30 cc nitrogen. Water is now added to the mixture in the flask so as to make the volume about 50 cc, and the liquid is heated on the

---

[1] A. Neufeld, Ann. **248**, 93.   [2] R. Overton, B. **26**, 10.
[3] H. Strache, M. **12**, 524; **13**, 299, Benedikt and Strache, *Ibid.* **14**, 270.

water-bath during 15-30 minutes; it is then cooled, diluted to the mark, well shaken, 50 cc transferred to the dropping funnel T, Fig. 8, and the determination

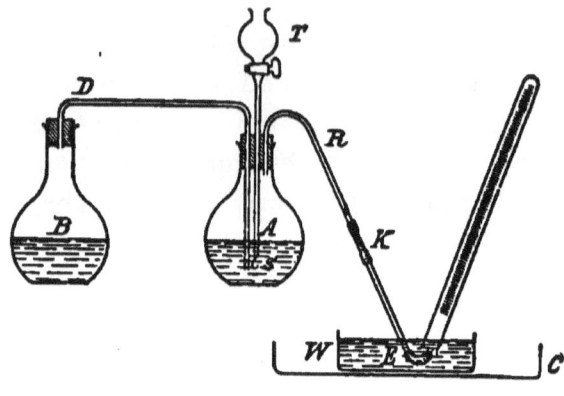

Fig. 8.

conducted in the manner described below. The flask A has a capacity of 750-1000 cc, and contains 200 cc of Fehling's solution, which is boiled whilst a rapid current of steam is blown in from the flask B. The tubes D and R must be flush with the rubber stoppers so as to promote the removal of air. The tube R is in two pieces, joined by the rubber tube K; its lower end is covered with a piece of rubber-tube E and dips below water in the dish W. The current of steam is continued until the bubbles of gas collected are very small, it is impossible, in a reasonable time, to remove all the air and a titration of the phenylhydrazine hydrochloride solution is made, previous to the actual determination, so as to allow for this error. 1 gram of the salt eliminates about 155 cc nitrogen, therefore, for the titration, 10 cc of the solution is accurately measured out, mixed with the needful proportion of sodium acetate solution, diluted to 100 cc, and 50 cc transferred to the dropping funnel; the end of this is drawn out at S and cut off at an angle so as to avoid the collection of bubbles of gas; before the funnel is fixed in place the stem is filled with water. When the greater portion of the air has been removed from the apparatus in the manner described above, the phenylhydrazine salt is allowed to mix with the Fehling's solution, care being taken to prevent the water flowing from W into A. When all has been added the funnel is washed out twice with hot water, which is, of course, also allowed to run into A. If the boiling is sufficiently brisk the evolution of nitrogen is completed in 2–3 minutes. As soon as the bubbles are as small as those of the air at the commencement of the experiment the heating is stopped, the hot water in W replaced by cold, the excess escaping into the dish C and the measuring tube removed to a cylinder of cold water. The actual determination is made immediately after the completion of the blank experiment and, if necessary, repeated a second or third time; since 200 cc Fehling's solution readily liberates 150 cc nitrogen, the quantity taken in A amply suffices for three or four carbonyl determinations.

As benzene is produced during the oxidation of the phenylhydrazine, a drop of it will be found floating on the surface of the water inside the measuring tube; this may be allowed for in measuring the gas or it may be removed. In the former case a little more benzene is introduced into the tube by means of a bent pipette, and, after remaining during a short time, the volume of nitrogen is read off in the ordinary manner; its reduction to 0° and 760 mm may be made by the help of the following table, the values in the second column being subtracted from the observed height of the barometer:

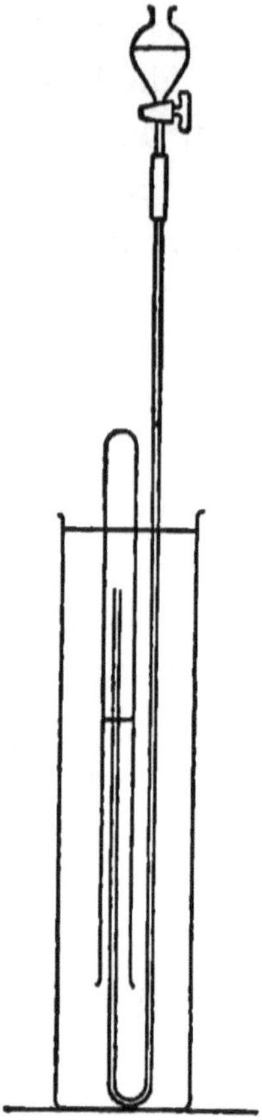

Temperature.	Tension of benzene + water.
15° C.	72.7 mm.
16	76.8
17	80.9
18	85.2
19	89.3
20	93.7
21	98.8
22	103.9
23	109.1
24	114.3
25	119.7

Fig 9.

The values given above are in part obtained by interpolation from Regnault's results and are therefore subject to error; for this reason, and on account of the high vapor tension of benzene its removal is advisable.[1] To accomplish this alcohol is added to the tube of nitrogen, which is placed in a cylinder of about its own length filled with water (Fig. 9). A glass tube 5 mm. in diameter is bent into the form of a U as shown in the figure, the smaller limb terminating in a jet and being of such length that, when the bent

---

[1] Benedikt and Strache, M. 14, 373.

portion rests on the bottom of the cylinder, the jet is several cm below the surface of the water. The longer limb rises about 40 cm above the surface of the water and is connected at the end by means of a piece of thick walled rubber tube with a dropping funnel. The U tube is completely filled with water and placed in the position shown in the figure. Alcohol (about 200 cc) is now allowed to flow from the funnel into the measuring tube; it issues from the jet in a fine stream and absorbs the benzene vapor present in the nitrogen as well as that floating on the water; the alcohol is removed in a similar manner by washing with at least 400 cc water, and the tube of nitrogen then removed to another cylinder of water, where, after a suitable interval, the volume of gas is read. The amount of carbonylic oxygen O is obtained from the volume of nitrogen, corrected to 0° and 760 mm, by the expression: $O = (g.V. - 2V_0.)$.

$$0.0012562 \cdot \frac{15.96}{28.02} \cdot \frac{100}{S}\% = O = (g.V. - 2V_0 \frac{0.07178}{S}\%,$$

where g is the weight of phenylhydrazine hydrochloride taken, V the volume of nitrogen evolved by 1 gram of this salt, S the weight of the compound employed, and V_0 the volume of nitrogen obtained at N. T. P. The theoretical value of V is 154.63cc, but the value employed in the calculation is that obtained in the blank experiment.

If the phenylhydrazone is in soluble in water or dilute alcohol, or if sparingly soluble phenylhydrazides are formed, the preparation of the phenylhydrazone must be made in alcoholic solution; in this case the weight of the column of liquid in the funnel T, Fig. 8, will not be sufficient to overcome the pressure of steam in the flask A. This difficulty may be surmounted by fitting the open end of the funnel with a rubber stopper, carrying a tube and stop-cock. By blowing through the tube the alcoholic liquor is forced into the flask, but great care is necessary, as the sudden evolution of alcoholic vapor may eject liquid from flask A to B, or may even lead to an explosion. A second objection to the use of alcohol is that, at its boiling-point, ketones do not always react quantitatively with phenylhydrazine. Both difficulties may be overcome by the use of recently boiled amylic alcohol as solvent, the portion of it which passes over with the nitrogen being subsequently removed simultaneously with the benzene, and in the same manner, by washing with alcohol and water.

(2) PREPARATION OF OXIMES.[1]

In the preparation of oximes the hydroxylamine is employed in the form of the *free base*, the *hydrochloride*, as *potassium hydroxylaminesulphonate*, or *zinc dihydroxylamine hydrochloride*. *Aldoximes* are obtained by treating aldehydes with an equimolecular proportion of hydroxylamine hydrochloride in concentrated aqueous solution, adding sodium carbonate (0.5 mol.), and allowing the mixture to remain at the ordinary temperature during $\frac{1}{2}$–8 days. The oxime is extracted with ether, the solution dried over calcium

---

[1] V. Meyer and Janny, B. 15, 1324, 1525. Janny, *Ibid.* 15, 2778; 16, 170.

chloride, and, after the removal of the ether, the residue rectified. An aqueous-alcoholic solution is used for aldehydes insoluble in water, and those that are readily oxidizable, such as benzaldehyde, are treated in flasks from which the air has been removed by means of carbonic anhydride.[1] Oximes of the carbohydrates, which are so readily soluble in water that they cannot be separated from the inorganic salts resulting from the use of hydroxylamine hydrochloride and sodium carbonate or sodium hydroxide, are treated with the calculated quantity of free hydroxylamine in alcoholic solution; after several days the oxime gradually crystallizes out.[2] Alcoholic solution of hydroxylamine is prepared by intimately mixing the hydrochloride with the necessary quantity of potassium hydroxide together with a little water, and then adding absolute alcohol, the clear liquid is afterwards separated from the precipitated potassium chloride.[3] The solution gradually acquires a slight yellow color,[4] which may be obviated by substituting sodium ethoxide for the potassium hydroxide.

Ketoximes are usually formed less readily than the aldoximes; for their preparation the ketone is mixed with sodium acetate and hydroxylamine hydrochloride in aqueous or alcoholic solution in the necessary proportions, and the liquid heated on the water-bath during 1–2 hours, or the ketone, in alcoholic solution, may be heated in a sealed tube with the hydrochloride at 160°–180° during 8–10 hours,[5] but sometimes in

---

[1] Petraczek, B. 15, 2783. [3] Wohl, *Ibid.* 24, 994. S., p. 367.
[2] Volhard, Ann. 253, 206. [4] Tiemann, B. 24, 994.
[5] Homolka, *Ibid.* 19, 1084.

these circumstances, instead of the oximes derivatives of them are formed by intramolecular rearrangement.[1] In many cases it is highly advantageous to allow the carbonyl derivative and the hydroxylamine to react in strongly alkaline solution; the proportions which usually give the best results are ketone, in alcoholic solution (1 mol), hydroxylamine hydrochloride (1.5–2 mol), alkaline hydroxide (4.5–6 mol); the last two are dissolved in the smallest requisite quantity of water.[2] The reaction is often completed at the ordinary temperature in a few hours; occasionally heating on the water-bath is desirable. This method cannot of course be used with ketones or aldehydes that are attacked by alkali, nor in the preparation of dioximes which readily change into their anhydrides in the presence of alkali. In such cases an acid liquid may be employed. Quinone furnishes an example of this. In alkaline solution it is reduced by hydroxylamine to hydroquinone, whilst in aqueous solution, in presence of hydrochloric acid and hydroxylamine hydrochloride, a dioxime is formed.[3] Some compounds, such as phenylglyoxalic acid, yield oximes both in alkaline and acid solution.[4] Oximes of ketonic acids may be obtained by treating the alkali salt in neutral aqueous solution with hydroxylamine hydrochloride; the precipitation of oxime usually commences at once, especially if the liquid is warmed.[5] Sometimes it is advisable to convert the acid into its methyl ester and

---

[1] Thorp, B. **26**, 1261.   [2] Auwers, *Ibid.* **22**, 609.
[3] Nietzki and Kehrmann, *Ibid.* **20**, 614.   [4] S. p. 370.
[5] Bamberger, *Ibid.* **19**, 1430.

avoid the use of excess of hydroxylamine hydrochloride so as to prevent the formation of nitriles.[1]

Potassium hydroxylamine sulphonate, supplied by the "Badischen Anilin- und Sodafabrik," under the name "Reducirsalzes," has been employed, in aqueous-alcoholic solution, for the preparation of oximes;[2] in presence of free alkali it is hydrolysed, and the liberated hydroxylamine acts in the nascent state on the carbonyl compounds.[3] It also possesses the advantage of cheapness.

Zinc dihydroxylamine hydrochloride, $ZnCl_2.2NH_2OH$, has been used chiefly for the preparation of ketoximes[4] as its resolution into hydroxylamine and anhydrous zinc chloride facilitates the elimination of water. It is prepared[5] by adding zinc oxide (1 part) to hydroxylamine hydrochloride (2 parts) in boiling alcoholic solution. The boiling is continued in a reflux apparatus for a few moments, and the liquid allowed to cool. The compound is deposited as a crystalline powder which dissolves sparingly in water or alcohol, but readily in solutions of hydroxylamine hydrochloride.

Ortho- and paraquinones and metadiketones cease to react with hydroxylamine if several atoms of hydogen in the ortho-position are replaced by haloïd atoms or alkyl groups.[6] Aromatic ketones of the formula $(CH_3.C)_2C.COR$, where R = phenyl or an

---

[1] Garelli, Gazz. 21, 2, *Ibid.* 2173.   [2] Kostanecki, B. 22, 1344.
[3] Raschig, Ann. 241, 187.
[4] Crismer, Bull. soc. chim. [3], 3, 114.   [5] B. 23, R. 223.
[6] Kehrmann, *Ibid.* 21, 3315.   Herzig and Zeisel, *Ibid.* 21, 3494. Cf. *Ibid.* 22, 1344.

alcohol radicle, are also incapable of forming oximes;[1] indeed, the presence of carbonyl, which does not yield oximes, in such compounds as acids,[2] amides,[3] or esters[4] may, by the production of hydroxamic acids, lead to erroneous results. The statement that alkyl salicylates and hydroxylamine give salicylhydroxamic acid requires further investigation.[4] The unsaturated ketoalcohol camphoroxalic acid

$$C_8H_{14}{<}{{C:C.OH.CO.OH}\atop{CO}}$$

yields an additive compound

$$C_8H_{14}{<}{{CH.C.OH.CO.OH}\atop{CO\ \ NH.OH}}$$

with hydroxylamine,[5] and it has been subsequently shown that certain unsaturated ketones, such as phorone, behave in a similar manner.[6]

(3) PREPARATION OF SEMICARBAZONES.[7]

The formation of well-crystallized derivatives of semicarbazide has proved extremely useful in the investigation of terpene compounds which often yield liquid oximes, and phenylhydrazones that only crys-

---

[1] V. Meyer, B. **29**, 836.   Feit and Davies, *Ibid.* **24**, 3546. Biginelli, Gazz. **24**, *I*, 437.   Claus, J. pr. **45**, 383.   Baum, B. **28**, 3209.
[2] Nef, Ann. **258**, 282.   [3] C. Hoffmann, B. **22**, 2854.
[4] Jeanrenaud, *Ibid.* **22**, 1273.
[5] Bishop Tingle, Am. Chem. Journ. **19**, 408.
[6] C. Harries and F. Lehmann, B. **30**, 231, 2726.
[7] Baeyer and Thiele, *Ibid.* **27**, 1918.

tallize with difficulty and readily undergo decomposition.

Preparation of Semicarbazide Salts.

(A) *Semicarbazide Hydrochloride*

$$NH_2.CO.NH.NH_2.HCl$$

is prepared from

(*a*) *Hydrazine sulphate*;[1]
(*b*) *Nitrocarbamide.*[2]

(*a*) Hydrazine sulphate (13 grams) is dissolved in water (100 cc), and neutralized with dry sodium carbonate (5.5 grams), when cold potassium cyanate (8.8 grams) is added and the solution allowed to remain overnight. A small quantity of hydrazodicarbonamide, $NH_2.CO.NH.NH.CO.NH_2$, s deposited which is somewhat augmented on acidifying with dilute sulphuric acid. The amide is removed and the acid liquid well shaken with benzaldehyde; the precipitate of benzalsemicarbazone which forms is separated and well washed with ether. It is now carefully heated on the water-bath, in portions of 20 grams, with concentrated hydrochloric acid (40 grams), sufficient water being added to cause the hot liquid to become clear; the benzaldehyde is removed by repeatedly extracting the hot liquid with benzene; when cold the aqueous solution deposits small needles of semicarbazide hydrochloride, which are removed, dried, and recrystallized from dilute alcohol. The purified compound forms prisms

---

[1] Thiele and O. Stange, B. **28**, 32.
[2] Thiele and Heuser, Ann. **288**, 312.

which decompose at 173°. The mother-liquors yield a further quantity of the benzal derivative when treated with benzaldehyde.

(b) Commercial nitrocarbamide (225 grams) is mixed with concentrated hydrochloric acid (1700 cc), a little ice added, and the liquid made into a paste by the successive additions of small quantities of zinc-dust and ice; constant stirring is necessary, and the temperature must not exceed 0°. The operation may be carried out in an enamelled dish cooled by means of a freezing mixture; when it is completed the product is allowed to remain for a short time, the excess of zinc-dust removed, and the filtrate saturated with sodium chloride. Sodium acetate (200 grams) is now added, together with acetone (100 grams), and the liquid placed on ice or in a freezing mixture. In the course of several hours a double salt of zinc chloride and acetone semicarbazide crystallizes out, it is collected and washed first with sodium chloride solution and finally with a little water. The yield is 40–55 per cent. The zinc compound (200 grams) is digested with concentrated ammonium hydrate (350 cc) and after some time the liquid is filtered; the residue consists of acetone semicarbazone, which is converted into semicarbazide salts in the manner described above for the benzal derivative. Many ketones do not readily react with semicarbazide hydrochloride and the products obtained from some may contain chlorine; in such cases semi-carbazide sulphate should be employed.

DETERMINATION OF CARBONYL. 77

(B) *Preparation of Semicarbazide Sulphate.*[1]

The filtrate from hydrazodicarbonamide, prepared in the manner described above, is cautiously made alkaline and shaken with acetone; the acetone semicarbazone which is deposited is mixed with alcohol, and treated with the calculated quantity of sulphuric acid; the sulphate crystallizes out and is purified by washing with alcohol.

Preparation of Semicarbazones.[2]

Semicarbazide hydrochloride, dissolved in the minimum quantity of water, is mixed with the calculated amount of potassium acetate in alcoholic solution, and the ketone added, together with water and alcohol sufficient to give a clear homogeneous liquid. This is allowed to remain until the completion of the reaction, which is recognized by the deposition of crystals when the mixture is diluted with water, and, as in the case of hydroxylamine, may require from a few minutes to four or five days. Sometimes it happens that the deposit produced is oily and only solidifies after several hours. The use of semicarbazide sulphate is illustrated by the preparation of ionone semicarbazone, which cannot be obtained from the hydrochloride. The sulphate is used in a finely divided form and added to glacial acetic acid in which the equivalent quantity of sodium acetate has been dissolved; after remaining at the ordinary temperature during twenty-four hours, the solution of ionone is

---

[1] Tiemann and Krüger, B. 28, 1754.   [2] Baeyer, *Ibid.* 27, 1918.

added, and the liquid allowed to remain three days longer. The product is poured into a considerable volume of water, extracted with ether, and the ether freed from acetic acid by treatment with sodium carbonate solution. After drying and removal of the ether the residue is treated with ligroin to remove some impurities, and the remaining product crystallized from a mixture of benzene and ligroin.

Should a ketone not yield a crystalline semicarbazide it is advisable to treat it with amidoguinadine picrate, as the resulting compounds are distinguished by the ease with which they crystallize.

(4) PREPARATION OF AMIDOGUINADINE DERIVATIVES.[1]

Preparation of Amidoguinadine Salts.[2]

Nitroguinadine (208 grams) is mixed with zinc-dust (700 grams) and sufficient ice and water to form a stiff paste; to this commercial glacial acetic acid (124 grams), diluted with its own volume of water, is added, the mixture is well stirred, and great care taken to add ice so that during the 2–3 minutes required for the addition of the acid the temperature shall not exceed 0°. The temperature is now allowed to rise gradually to 70°; at this stage the mixture is viscid and has a yellow color due to an intermediate product. The temperature is maintained at 40°–45° until a little of the filtered liquid ceases to yield a red color with sodium hydroxide and a ferrous salt. The conclusion of the operation is usually in-

---

[1] Baeyer, B. **27**, 1919.   [2] Thiele, Ann. **270**, 23.

dicated by evolution of gas and the formation of a frothy scum on the surface of the liquid. The product is filtered, the residue well washed with water, the washings and filtrate mixed with hydrochloric acid sufficient to liberate the acetic acid, and the whole concentrated to the smallest possible bulk; it is then treated with alcohol, again evaporated to expel water, and the solid boiled out with alcohol; this, when cold, deposits amidoguinadine hydrochloride, which is further purified by recrystallization from alcohol to which animal charcoal has been added. The pure salt melts at 163°.

Preparation of Amidoguinadine Bicarbonate.[1]

The liquid obtained by the reduction of nitroguinadine with zinc-dust and acetic acid is maintained slightly acid with acetic acid, evaporated to about 500 cc, cooled, and treated with concentrated sodium or potassium bicarbonate solution to which a little ammonium chloride has been added to prevent the deposition of any zinc. The amidoguinadine salt is completely precipitated in twenty-four hours; it is sparingly soluble in hot water but suffers decomposition, and when slowly heated it melts and decomposes at 172°.

The nitrate and the normal and hydrogen sulphates are prepared in a similar manner.

---

[1] Thiele, Ann. **302**, 333.

Preparation of Amidoguinadine Picrate Derivatives.

Amidoguinadine hydrochloride is dissolved in a small quantity of water containing a trace of hydrochloric acid and the ketone added together with sufficient alcohol to give a clear solution. The reaction is completed by boiling for a short time. The product is treated with water and sodium hydroxide solution in excess, and the liquid base extracted by means of ether. The ethereal solution is separated, the ether removed, the residual oil suspended in water and treated with picric acid in aqueous solution, the picrate is quickly deposited in granular crystals which are purified by recrystallization from concentrated or dilute alcohol.

Some carbohydrate derivatives of amidoguinadine are known.[1]

PARAMIDODIMETHYLANILINE DERIVATIVES.

Condensation products of aldehydes and paramidodimethyl aniline may be prepared by mixing the constituents with or without the addition of alcohol. The temperature of the liquid rises spontaneously, and the condensation product usually separates in crystals.[2]

---

[1] Wolff and Herzfeld, Z. Rüb. *1895*, 743. Wolff, B. **27**, 971; **28**, 2613.

[2] A. Cahn, *Ibid.* **17**, 2938. The literature of this subject is given in M. and J. II, p. 515.

Chapter IV.

DETERMINATION OF THE AMINO $\overset{\text{\tiny I}}{\text{N}}H_2$; NITRILE $\overset{\text{\tiny I}}{\text{C}}$N; AMIDE $\overset{\text{\tiny I}}{\text{C}}$O.NH$_2$; IMIDE $\overset{\text{\tiny II}}{\text{N}}$H; METHYL IMIDE $\overset{\text{\tiny II}}{\text{N}}$.CH$_3$; AND ETHYL IMIDE $\overset{\text{\tiny II}}{\text{N}}$.C$_2H_5$ GROUPS.

DETERMINATION OF THE AMINO GROUP ($\overset{\text{\tiny I}}{\text{N}}H_2$).

Different methods are employed for the determination of the amino group according to whether the compound is an aromatic or aliphatic amine.

(A) Determination of Aliphatic Amino Groups.

These are determined:

(1) *By means of nitrous acid.*
(2) *By analysis of the salts and double salts.*
(3) *By acetylation.*

(1) *Nitrous Acid Method.*—Aliphatic amines react with nitrous acid in accordance with the equation

$$\text{RNH}_2 + \text{HNO}_2 \longrightarrow \text{ROH} + \text{N}_2 + \text{H}_2\text{O}.$$

The first method suggested for the determination of the nitrogen consisted in liberating it in an atmosphere of nitric oxide, which was then absorbed by means of ferrous sulphate solution.[1] The following process is much more convenient. The substance, dissolved in

---

[1] R. Sachsse and W. Kormann, Landwirthsch. Vers.-Stationen, 17, 321. Z. anal. Ch. 14, 380.

just sufficient dilute sulphuric acid to give a neutral solution, is placed in a flask provided with a trebly bored stopper. If possible a distillation bulb should be employed having a capillary tube fused to it. A dropping funnel is fitted to the stopper of the flask, the leg being drawn out, bent upwards, and passed below the surface of the liquid; it is filled with distilled water at the commencement of the experiment. The third tube of the flask, or the side tube of the distillation bulb, is fitted by means of an airtight stopper almost to the bottom of a second distillation-bulb. This has its side tube suitably bent, and connected with a Liebig's potash bulb filled with potassium permanganate solution (3 per cent) containing sodium hydroxide (about 1 gram). The gas delivery-tube is attached to the potash bulbs and dips below the mouth of the measuring vessel, which is half filled with mercury and half with potassium hydroxide (sp. gr. = 1.4). The air is displaced from the apparatus by a slow current of carbonic anhydride, which may be obtained pure and free from air by dropping dilute sulphuric acid (50 per cent. sp. gr. = 1.4) into a concentrated solution of potassium carbonate (sp. gr. = 1.45–1.5).[1] When the air is expelled the measuring tube is placed in position and a slight excess of potassium nitrite solution added by means of the dropping-funnel. The reaction is completed by heating on the water-bath and the addition of a little dilute sulphuric acid.

(2) *Analysis of Salts and Double Salts.*—The prep-

---

[1] Fr. Blau, M. **13**, 280.

aration of most of these is too well known to require comment. Of the simple salts the *hydrochlorides* sometimes can only be induced to crystallize in a state of purity by the action of anhydrous hydrogen chloride on a solution of the base in ether free from alcohol and moisture. The *chromate* and *picrate*,[1] especially the latter, usually crystallize readily. The *mercurichloride*, $RHgCl_3$, has occasionally been of service in cases where the *auro-chloride* or *platino-chloride* are oily or unstable (cf. pp. 89, 94).

(3) *Acetylation.*—This is described in the next section on the acetylation of the aromatic amines.

(B) Determination of Aromatic Amino Groups.

The following methods are employed for the determination of primary aromatic amines:

(1) *Titration of the salts.*
(2) *Preparation of diazo derivatives.*
   (a) *By conversion into an azo dye.*
   (b) *Indirect method.*
   (c) *Azoimide method.*
   (d) *By means of the Sandmeyer-Gattermann reaction.*
(3) *Analysis of salts and double salts.*
(4) *Acetylation.*

(1) TITRATION OF THE SALTS.[2]

(I) Salts of aromatic amines, in aqueous or alcoholic solution, give an acid reaction with rosolic acid or

---

[1] Delépine, Bull. 15, 53.
[2] Menschutkin, B. 16, 316.

phenolphthaleïn. The salt, preferably the hydrochloride or sulphate, is dissolved in water or dilute alcohol, phenolphthaleïn added, and the titration carried out in the ordinary manner with potassium hydroxide.

(II) Many free bases may be directly titrated with hydrochloric acid, methyl orange being used as an indicator.

(2) PREPARATION OF DIAZO-DERIVATIVES.

(a) *Conversion of the Base into an Azo Dye.*[1]

The base, for example aniline (0.7–0.8 gram), is dissolved in hydrochloric acid (3 cc) and diluted with water and ice to 100 cc. A titrated solution of "R-salt," sodium 2 : 3 : 6 naphtholdisulphonate, is prepared, of such strength that a liter is equivalent to about 10 grams of naphthol. The solution of the hydrochloride is cooled to 0°, sodium nitrite added in quantity equivalent to the aniline or other base present, and the mixture gradually poured into a measured quantity of the sulphonate solution which has been treated with sodium carbonate in excess. The dye produced is precipitated by means of sodium chloride, filtered, and the filtrate tested with diazobenzene chloride solution and R-salt to determine whether the latter or the base is in excess. By repeating the experiment it is possible to find the volume of R-salt solution necessary to combine with the diazo-derivative of the base originally taken.

---

[1] Reverdin and De la Harpe, Ch. Ztg. **13**, I. 387, 407; B. **22**, 1004.

The following method has been applied to aniline, ortho- and paratoluidine, metaxylidine, and sulphanilic acid.[1] A known quantity of the base is diazotized and made made up to a certain volume; it is then immediately added from a burette to a solution of "Schäfer's salt," sodium 2 : 6-naphthol sulphonate, of known strength, which has been mixed with sodium chloride and a few drops of ammonium hydroxide, the addition being continued so long as a precipitate forms. The end point is determined by bringing a drop of the clear supernatant liquor on to filter paper and allowing it to come into contact with a drop of the diazo-solution. The progress of the reaction can be followed by the intensity of the red color produced at the point of contact of the two liquids on the paper. Towards the end of the operation the color is only visible in the middle of the moist circle. In the case of readily soluble dyes, such as that given by sulphanilic acid, the paper must be covered with a thin crust of sodium chloride and the test portions allowed to fall on to it; more sodium chloride must also be added to the naphtholsulphonate solution.

(b) *Indirect Method.*

This is extensively employed for technical purposes and consists of an inversion of a method for the determination of nitrous acid.[2] The base is treated with three times its weight of hydrochloric acid and the mixture dissolved in so much water that the solution

---

[1] R. Hirsch, B. **24**, 324.
[2] A. G. Green and S. Rideal, Ch. N. **49**, 173.

contains 0.01 to 0.1 gram equivalent of the base. The solution is maintained at 0° by means of ice, and titrated with sodium nitrite solution, potassium iodo-starch paper being used as indicator; the operation is ended when a drop of the mixed liquids gives a blue coloration with the paper. The nitrite solution should be about $N/10$. It is prepared[1] by dissolving the nitrite in 300 parts of cold water and is titrated by adding $N/10$ potassium permanganate solution until a distinct permanent red coloration is obtained; two or three drops of dilute sulphuric acid are now added, then immediately excess of the permanganate, the liquid is made strongly acid with sulphuric acid, heated to boiling, and the excess of permanganate determined by means of $N/10$ oxalic acid solution.

(c) *Azoimide Method.*[2]

This is specially applicable to compounds containing amino groups linked to different nuclei. The azoimides are prepared by the action of ammonia on the diazoperbromides[3] and, on account of the large content of nitrogen in the former, their analysis is peculiarly well adapted for the determination of the number of diazotisable groups in the molecule. Details of the method of preparing azoimides have been given by various chemists[4].

---

[1] L. P. Kinnicutt and J. U. Nef, Am. Chem. Journ. **5**, 388. Fresenius' Zschr. **25**, 223.
[2] Meldola and Hawkins, Ch. N. **66**, 33.
[3] Griess, Ann. **137**, 65.
[4] Nölting, Grandmougin, and O. Michel, B. **25**, 3328. Curtius and Dedichen, J. pr. [2], **50**, 250.

(d) *Sandmeyer's[1]-Gattermann's[2] Reaction.*

The determination of the amino group is often conveniently accomplished by converting it into the diazo-derivative and replacing the nitrogen by chlorine; as a rule the diazo-compound is not isolated. The following example[3] will serve to illustrate the method: Metanitraniline (4 grams) and concentrated hydrochloric acid, sp. gr. = 1.17 (7 grams), are dissolved in water (100 grams), and 10 per cent cuprous chloride solution (20 grams) added; the mixture is heated almost to boiling in a reflux apparatus, and sodium nitrite (2.5 grams), dissolved in water (20 grams) is gradually run in by means of a dropping funnel, the mixture being well shaken during the addition. Nitrogen is evolved and a heavy brown oil collects which solidifies when cooled with ice and is purified by distillation. As a rule these chloro-derivatives are volatile with steam; if not they are purified by means of ether or benzene.

The above method is the one originally proposed by Sandmeyer; by means of it chloro-compounds may be readily obtained from diamines which cannot be diazotised in the ordinary manner. The cuprous chloride employed is prepared by boiling crystallized copper sulphate (25 parts) and anhydrous sodium chloride (12 parts) with water (50 parts); some sodium sulphate crystallizes out, and when the reaction is completed the product is mixed with concentrated hydrochloric acid (100 parts) and copper turnings

---

[1] B. 17, 1633.   [2] *Ibid.* 23, 1218.   [3] *Ibid.* 17, 2650.

(13 parts), the mouth of the flask is loosely closed and the mixture boiled until the liquid becomes colorless. Sufficient concentrated hydrochloric acid is now added to bring the weight of the mixture to 203.6 parts, since only 6.4 parts of the copper actually dissolve, 197 parts of solution are obtained which contains 0.2 gram molecules of Cu. Cl. The filtered solution may be retained a considerable time in a well closed bottle containing carbonic anhydride.[1]

Cupric chloride is reduced to cuprous chloride by hypophosphorus acid,[2] hence, in place of the cuprous chloride solution prepared according to the foregoing method, a mixture of hydrochloric acid, copper sulphate solution, and sodium hypophosphite may be employed.[3]

The use of finely divided copper instead of cuprous chloride has been suggested;[4] amongst other advantages the reaction proceeds at the ordinary temperature, and the yield is frequently improved. The copper is prepared by adding zinc-dust, through a fine sieve, to a cold saturated solution of copper sulphate until only a faint blue color remains, the product is well washed by decantation with large quantities of water, the remaining zinc removed by digestion with highly dilute hydrochloric acid, and the copper filtered and washed with water until neutral; it is preserved in the form of a paste in well-closed bottles. The following example will illustrate the method of working: Aniline (3.1 grams) is mixed with 40 per cent.

---

[1] Feitler, J. pr. **4**, 68.   [2] A. Cavazzi, Gazz. **16**, 167.
[3] A. Angell, *Ibid.* **21**, *2*, 258.   [4] Gattermann, B. **23**, 1218.

hydrochloric acid (30 grams) and water (15 cc), the liquid is cooled to 0° and a saturated aqueous solution of sodium nitrite (2.3 grams) quickly added, the liquid being vigorously stirred, preferably by means of a turbine; the reaction is completed in one minute. Finely divided copper (4 grams) is now gradually added to the diazo solution, which is well stirred; the reaction requires 15–30 minutes for completion, this is signalized by the particles of copper ceasing to be carried to the surface of the liquid by the escaping bubbles of nitrogen. The chlorobenzene is removed by steam distillation.

(3) ANALYSIS OF SALTS AND DOUBLE SALTS.

The remarks on the salts of aliphatic amines (p. 83) apply generally to those of the aromatic series; the accumulation of negative groups in their molecules often completely prevents the formation of salts. As a rule the *auro-chloride* contains one atom of gold for each amino group, and the *platino-chloride* one atom of platinum to two amino groups, but amino pyridine platino-chloride has the formula $(C_5H_6N_2)_3.H_2PtCl_6$.[1] Sometimes the *alkyl haloïd salts* are of service, but many primary bases do not form them.[2] In presence of secondary or tertiary amino groups the method yields fallacious results. Certain compounds free from nitrogen may form salts, dimethylpyrone, for example, gives amongst others a platino-chloride $(C_7H_8O_2)_2.H_2PtCl_6$.[3]

---

[1] M. 15, 176.   [2] Hofmann, Jahresbericht (*1863*), p. 421.
[3] Collie & Tickle, Journ. Chem. Soc. 75, 712.

(4) ACETYLATION.

The methods of acetylation described for the determination of hydroxyl (Chapter I) are also applicable to the amino group (cf. acetylation of imides, p. 92).

DETERMINATION OF THE NITRILE GROUP (C:N.)

The nitrile radicle is determined by hydrolysis, the resulting ammonia or acid being collected.

(*a*) Prolonged boiling with hydrochloric acid is usually sufficient to cause hydrolysis; the product is then treated with alkali in excess, and the ammonia distilled off and determined in the ordinary manner.

(*b*) Should the hydrolysis only take place in the presence of aqueous or alcoholic alkali an apparatus similar to that employed in Zeisel's method for the determination of methoxyl is used (Fig. 1, p. 34). A current of air, freed from carbonic anhydride, is passed through the apparatus, and the bulbs are filled with concentrated alkali solution; the ammonia is most readily determined as the platino-chloride. At the conclusion of the experiment the flask A will contain the alkali salt of the acid produced, and may be treated by one of the methods described for the determination of carboxyl (Chapter II).

(*c*) The hydrolysis of nitriles [1] may be hindered by stereo-chemical influences, especially in the case of diortho-substituted compounds,[2] just as the corresponding acids etherify with difficulty, or not at all, under the influence of hydrogen chloride. The nitriles

---

[1] M. and J., II, p. 545.

[2] A. W. v. Hofmann, B. **17**, 1914; **18**, 1825; Stallburg, Ann. **278**, 209. Cain, B. **28**, 969. V. Meyer and Erb, *Ibid.* **29**, 834, footnote. Sudborough, Journ. Chem. Soc. **67**, 601.

in question, although they resist prolonged heating at a high temperature in a sealed tube with hydrochloric acid, are all converted into amides by continued boiling with alcoholic potassium hydroxide.[1] The amide is hydrolysed to the acid in the manner described in the following section: Cyanmesitylene[2] requires boiling during seventy-two hours with alcoholic potassium hydroxide, and triphenylacetonitrile[3] needs fifty hours boiling with the same reagent to produce the amide. Some nitriles that are otherwise resistant may be hydrolysed by heating at $120°$–$130°$ during an hour with 90 per cent sulphuric acid (20–30 parts). The resulting amide is converted into the acid by means of nitrous acid[4] (cf. following section). Unhydrolysable nitriles have also been described.[5]

(*d*) Certain amides may be obtained by the action of alkaline hydrogen peroxide at $40°$[6] on the nitriles; the resulting compounds are then treated in the manner described in the preceding section.

DETERMINATION OF THE AMIDO GROUP ($\overset{1}{C}O.NH_2$).

The amido group is determined by hydrolysis, in a similar manner to the nitrile group (preceding section). The method employed for the hydrolysis of very stable amides[7] is best illustrated by its application to the preparation of triphenylacetic acid.[8] The

---

[1] Bouveault. S. p. 80. Hantzsch and Lucas, B. **28**, 748.
[2] V. Meyer and Erb, *Ibid.* **29**, 834. [3] V. Meyer, *Ibid.* **28**, 2782.
[4] Sudborough, Jour. Chem. Soc. **67**, 601. Münch, B. **29**, 64.
[5] Radziszewski, *Ibid.* **18**, 355.
[6] Claus and Wallbaum J. pr. **56**, 52.
[7] Bouveault, Bull. [3], **9**, 370.
[8] G. Heyl and V. Meyer, B. **28**, 2783.

finely divided amide (0.2 gram) is gently warmed with concentrated sulphuric acid (1 gram) and the clear solution cooled in ice, sodium nitrite (0.2 gram), dissolved in water (1 gram) cooled to 0° is added very slowly by means of a capillary tube; when the addition is complete the test-tube containing the mixture is placed in a beaker of water and gradually heated. The evolution of nitrogen commences at 60°–70°, and is completed at 80°–90°; finally the tube is heated in boiling water for 3–4 minutes, but not longer. When cool, ice is added to the liquid, the precipitated solid collected, and purified by solution in dilute sodium hydroxide and precipitation with sulphuric acid. It is highly desirable to use the exact theoretical quantity of sodium nitrite dissolved in the smallest possible volume of water.[1] Stereo-chemical influences are effective in hindering the hydrolysis of amides as they are that of the nitriles.[2]

DETERMINATION OF THE IMIDE GROUP ($\overset{\shortmid\shortmid}{N}H$).

The following methods are employed for the determination of the imide group:

(1) Acetylation.
(2) Alkylation.
(3) Analysis of salts.
(4) Elimination of the imidogen as ammonia.

(1) ACETYLATION OF IMIDES (SECONDARY AMINES).

Imides may be acetylated by any of the methods employed for the determination of hydroxyl which are

---

[1] Sudborough, Jour. Chem. Soc. **67**, 604.
[2] A bibliography of the subject is given in M. and J. II, p. 545.

described in Chapter I. The reaction usually takes place without difficulty, and therefore an indirect method[1] may be utilized. A weighed quantity of the compound (about 1 gram) is placed in a flask, fitted to a reflux apparatus, and acetic anhydride (about 2 grams) quickly added. The anhydride should be added from a suitably stoppered vessel, which is weighed before and after the addition. The mixture is allowed to remain at the ordinary temperature during about thirty minutes, water (50 cc) is then added, and the liquid heated on the water-bath during forty-five minutes; the solution is now cooled, diluted to a definite volume, and titrated with sodium hydroxide of known strength, phenolphthaleïn being used as indicator.

The process was specially worked out for methylaniline, hence, for other imides, the duration of the heating and the temperature require modification according to the readiness with which they react. It may be desirable to heat in a sealed tube, or in a dry closed flask, the mixture being constantly shaken, and the anhydride diluted with ten volumes of dimethylaniline.[2]

(2) ALKYLATION OF IMIDES.

Some imide groups may be methylated by dissolving the compound in alkali and gradually adding methylic iodide; the mixture is constantly shaken and maintained at the ordinary temperature. The method has been extensively employed in the investigation of purin and uric acid derivatives.[3]

---

[1] Reverdin and De la Harpe, B. **22**, 1005.
[2] H. Giraud, Bull. (3), II. 142.
[3] E. Fischer, B. **28**, 2479; **30**, 569, 3094; **32**, 453. C. (1897), II. 157.

(3) ANALYSIS OF SALTS.

The remarks on the analysis of salts of primary amines (pp. 83, 89) apply equally to those of secondary ones.

(4) ELIMINATION OF IMIDOGEN AS AMMONIA.

The hydrolysis of the imides is usually carried out by prolonged boiling with hydrochloric acid either in an open vessel or under pressure in a sealed tube. The liquid is then made alkaline, the ammonia or amine volatilized into hydrochloric acid, and the excess of the latter determined by titration or, in some cases, by means of the platino-chloride.

DETERMINATION OF METHYL IMIDE ($\overset{\shortparallel}{N}CH_3$).[1]

The hydriodides of methylated bases eliminate methyl iodide at 200°–300° in accordance with the equation $R_2NCH_3 \cdot HI \longrightarrow R_2NH + CH_3I$; the iodide may be determined by Zeisel's method (cf. p. 33).

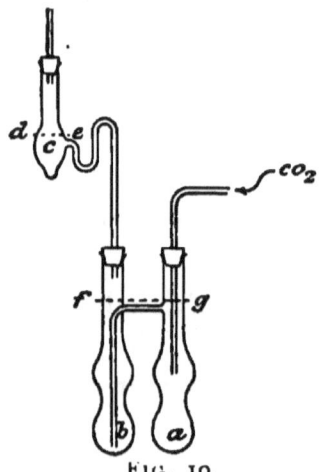

FIG. 10.

The apparatus employed is identical with that of Zeisel except the vessel in which the substance is heated. This is shown in Fig. 10, and consists of a double flask $a\,b$ connected by means of a cork with the vessel c. The method is modified according to whether one or more alkyl groups are linked to nitrogen, and, in the latter case, whether these are to be

---

[1] J. Herzig and H. Meyer, B. **27**, 319. M. **15**, 613; **16**, 599.

DETERMINATION OF THE AMINO GROUP, ETC. 95

determined successively; finally the presence of alkyloxy groups, in addition to methyl imide, demands special manipulation.

(1) *Determination with only one Alkyl linked to Nitrogen.*

The compound (0.15–0.3 gram) as free base, nitrate, or haloïd salt is placed in the flask *a* together with sufficient hydriodic acid (sp. gr. = 1.68–1.72) to fill the vessel *c* to the mark *de*; the object of this is to retain any volatile basic compounds which might be carried over by the carbonic anhydride. In addition to the acid the flask *a* also contains ammonium iodide in quantity equal to 5–6 times that of the substance employed. The vessel *C* is connected directly with the condenser (Fig. 1, p. 34); it should contain a little red phosphorus if much iodine is liberated in *a*, as is usually the case when nitrates are employed. The flask *b* is filled with asbestos, a little of which is also placed in *a* to facilitate the boiling. A more rapid current of carbonic anhydride is used than in the determination of methoxyl, so as to remove the methyl iodide as quickly as possible and prevent its entering into combination with the other compounds produced, consequently two absorption flasks with silver nitrate must always be employed. The heating is done by means of a sand-bath of copper with a sheet-iron bottom; it is divided into two equal portions by a partition, and is of such a shape as to permit the flasks being immersed in the sand up to the line *fg*. The flask *a* is first heated, carbonic anhydride being passed through the apparatus, a portion of the acid distils into *b* and some into *c*. Gradually the second chamber of the bath is filled with sand, and *b* then directly heated. All the acid soon accumulates in *c*, the carbonic anhydride bubbling through it whilst the flask *a* contains only the hydriodide of the base. The commencement of the decomposition is indicated by a turbidity in the silver nitrate solution, and it occurs soon after the acid has been expelled from the flask *b*. The remainder of the experiment is carried out exactly as in the methoxyl determination.

(2) *Determination with two or more Alkyl Groups linked to Nitrogen.*

This is carried out in the manner described in the preceding section; when the operation is completed the appararus is allowed to cool in a current of carbonic anhydride, *c* is detached from the condenser, and by cautious tilting the acid poured from it back to *b*, whence it will pass spontaneously to *a*. A fresh quantity of silver nitrate is placed in the absorption flasks, and the apparatus is ready to heat again. The operation is repeated until the quantity of silver iodide obtained is equivalent to an amount of alkyl weighing less than 0.5 per cent of the substance employed. It is important to conduct the determinations at the lowest possible temperature, and therefore a thermometer is placed in the sand-bath which is never allowed to exceed, by more than 40°, the temperature (200°–250°) at which the silver nitrate solution first becomes turbid. When several alkyl groups are present, it is advisable to use more ammonium iodide than otherwise, about 5 grams in *a*, and 2–3 grams in *b*. Each decomposition requires some two hours for completion, and three such are amply sufficient even though the compound contains three or four alkyls.

(3) *Successive Determination of the Alkyl Groups.*

The alkyl groups may be successively eliminated from feebly basic compounds such as caffeine or theobromine. In place of the vessel previously employed (Fig. 10), the substance is heated in one of the shape shown in Fig. 11. It is immersed in a sand-bath to the mark *ab*; after heating the acid is allowed to flow back to the flask, a little ammonium iodide is added, and the heating repeated,—the operation being performed a third time, with the addition of more ammonium iodide, if three alkyl groups are present.

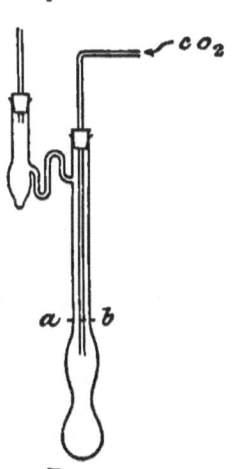

FIG. 11.

(4) *Determination of Methyl Imide in Presence of Methoxyl.*

The methyl imide may be determined in presence of methoxyl by heating the hydriodide alone in the flask *a* (Fig. 10); it is, however, preferable to add to it hydriodic acid (10 cc), and heat the flask in an oil- or glycerol-bath so that scarcely any distils over into *b*. When the operation is ended, which is indicated by the silver nitrate solution becoming clear, the temperature is raised, and the acid distilled off until only so much remains in *a* as is usually employed for the methyl imide determination (see section 1). During the distillation the silver nitrate solution remains quite clear, and the methoxyl determination is completed. A fresh portion of silver nitrate is taken, the excess of acid removed from *b* and *c*, ammonium iodide added, and the methyl imide determination commenced in the manner described in the preceding sections.

(5) *General Remarks on the Method.*

The purity of the hydriodic acid and ammonium iodide must be ascertained by means of a blank experiment.

The method is applicable to all compounds which can form a hydriodide even though this may not be capable of isolation, and accurate results are obtained by the use of the hydrobromide, hydrochloride, or nitrate. Quantitative results are also obtained in the case of many compounds, such as *n*-ethylpyrroline, methylcarbazole, and dimethylparabanic acid, which do not form salts. The limits of error lie between $+3$ and -15 per cent of the total alkyl, consequently the presence or absence of one such group can only be determined with certainty when the theoretical difference in composition for one alkyl exceeds 2 per cent, or, in other words, when the molecular weight of the original methylated compound is less than 650.

In considering the results obtained it is necessary to observe the color of the silver iodide; should this be dark or gray instead of yellow, the error is almost always positive.

$$100 \text{ parts AgI} = 6.38 \text{ parts of } CH_3.$$

DETERMINATION OF ETHYL IMIDE ($\overset{\text{II}}{N}C_2H_5$).

The method[1] of determination is exactly the same as that described above for methyl imide. 100 parts Ag I = 12.34 parts of C_2H_5.

DIFFERENTIATION OF THE METHYL IMIDE AND ETHYL IMIDE GROUPS.

The method of determination by means of the alkyl iodides does not, as a rule, distinguish between ethyl imide and methyl imide; in doubtful cases it is necessary to distil a considerable quantity of the hydriodide of the base, and purify and identify the alkyl iodide which is evolved. A second method consists in distilling the base with potassium hydroxide, evaporating the distillate with hydrochloric acid to dryness, separating the organic hydrochlorides from ammonium chloride by means of absolute alcohol, and converting the former into picrates, platinochlorides, etc., which may then be identified; the method must, however, be used with caution, as it may lead to erroneous results.

---

[1] J. Herzig and H. Meyer, B. 27, 319. M. 15, 613; 16, 599.
[2] Ciamician and Boeris, B. 29, 2474.

Chapter V.

DETERMINATION OF THE DIAZO GROUP (R.N:N.R); OF THE HYDRAZIDE RADICLE ($\overset{|}{N}H.NH_2$); OF THE NITRO-GROUP (NO_2); OF THE IODOSO-GROUP (IO); OF THE IODOXY-GROUP (IO_2); OF THE PEROXIDE GROUP $\overset{\scriptscriptstyle\|}{C}\!\!<\!\!\overset{O}{\underset{O}{|}}$; IODINE NUMBER.

DETERMINATION OF THE DIAZO GROUP (R.N:N.R).

The aliphatic and aromatic diazo-compounds are differently constituted, hence the methods adapted for their determination are not identical.

(A) Aliphatic Diazo-compounds $\left(C-CH\!\!<\!\!\overset{N}{\underset{N}{\|}}\right)$.

The following methods are employed:[1]

(1) *Titration with iodine.*
(2) *Analysis of the iodo-derivatives.*
(3) *Determination of the nitrogen in the wet way.*

(1) DETERMINATION OF THE NITROGEN BY TITRATION WITH IODINE.

This reaction takes place in accordance with the equation $CHN_2.COOR + I_2 \longrightarrow CHI_2.COOR + N_2$.

Rather more than the theoretical quantity of iodine is accurately weighed, dissolved in absolute ether, and added, by means of a burette, to a known quantity

---

[1] Curtius, J. pr. **146**, 422.

of the diazo-compound also in ethereal solution; the end of the reaction is indicated by a sharp change in the color of the diazo-compound from lemon yellow to red; towards the conclusion of the titration the reaction is facilitated by warming the liquid on the water-bath. The excess of iodine solution is run into a tared flask, the ether cautiously removed, and the residue weighed. Unless the compound employed is in a high state of purity the change of color in the liquid takes place long before all the nitrogen has been expelled.

(2) ANALYSIS OF THE IODINE DERIVATIVE.

The iodine in the iodo-compound may be determined in the ordinary manner, or the following simpler method, first used in the investigation of diazoacetamide,[1] may be employed. A weighed quantity of the substance is placed in a tared beaker, dissolved in a little absolute alcohol, and iodine added until a permanent red coloration is obtained. The alcohol is volatilized on the water-bath, the excess of iodine removed by cautious heating, and the crystalline residue weighed. In this case also the compound employed must be pure.

(3) DETERMINATION OF THE NITROGEN IN THE
WET WAY.

On account of the great volatility of the aliphatic ethereal diazocarboxylates the method of nitrogen

---

[1] Curtius, J. pr, **146**, 423.

determination described on p. 81 cannot be employed. This difficulty is overcome[1] by the use of the apparatus shown in Fig. 12.

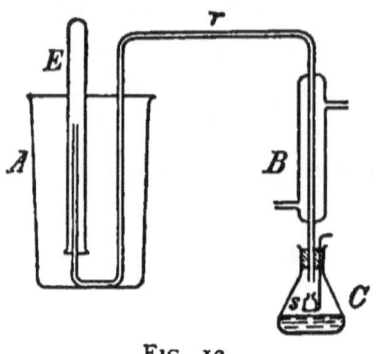

FIG. 12.

A is a large gas cylinder containing water, r a capillary tube, the upper open end of which rises a little above the level of the water in A. E is a gas measuring tube, B a small condenser fitted to the little flask C by means of a rubber stopper; through this a platinum wire also passes. It is bent in the manner shown and carries a glass-stoppered vessel such as is employed in vapor density determinations. The flask C is partially filled with well boiled, highly dilute sulphuric acid, the compound (about 0.2 gram) weighed into the small vessel s, and the apparatus fitted together air-tight. When the air in the apparatus is in equilibrium with the atmosphere, which can readily be observed if a drop of water is placed in r, the volume of air in the eudiometer tube is read off, and the temperature noted. The vessel s is now dropped into the acid, which is gradually heated to boiling; the decomposition is completed in a few minutes. The apparatus is allowed to cool completely, the level of water in and outside the tube E adjusted, and the volume, temperature, and pressure noted; the difference in volume from the previous reading gives the quantity of nitrogen evolved. As a

---

[1] Curtius, J. pr. **146**, 417.

rule the pressure does not materially change during the experiment.

Compounds containing an amino as well as a diazo-group, such as diazoacetamide, may be decomposed by means of dilute hydrochloric acid; after the evolution of nitrogen is completed the ammonium chloride in the flask c may be precipitated with platinum chloride and the amido and diazo nitrogen thus separately determined in one operation.

(B) Aromatic Diazo Compounds. (Diazonium Derivatives $C.N.OH$)
$$\overset{\text{III}}{N}$$

The diazo group in aromatic compounds is usually determined by the preceding method[1] (3), but it is preferable to employ a Lunge's nitrometer and 40 per cent sulphuric acid.[2] If the compound is unstable and the determination is made in a current of carbonic anhydride[3] the air should be expelled at a temperature of 0°.[4] Sulphuric acid, sp. gr. = 1.306, has a vapor tension of 9.4 mm at 15°[5].

DETERMINATION OF THE HYDRAZIDE GROUP ($\overset{\text{I}}{N}H.NH_2$).

Either the oxidation or iodometric method may be employed.

---

[1] Knoevenagel, B. 23, 2997. v. Pechmann and Frobenius, *Ibid*. 27, 706.
[2] Bamberger, *Ibid*. 27, 2598.
[3] H. Goldschmidt and A. Merz, *Ibid*, 29, 1369; 30, 671.
[4] Hantzsch, *Ibid*. 28, 1741.   [5] Regnault.

(1) OXIDATION OF HYDRAZIDES.[1]

Boiling Fehling's solution hydrolyses acid hydrazides, and oxidizes the resulting phenyl hydrazine, the nitrogen of which is evolved quantitatively and determined by the method described on p. 65. The compound is dissolved if possible in water or alcohol; hydrazides which do not dissolve are weighed into a small stoppered vessel which is fixed mouth downwards in the hole of the stopper otherwise occupied by the funnel A, Fig. 8, and is dropped into the boiling solution by means of a glass rod of the same volume. Insoluble compounds may also be treated according to the following method:[2] 100 cc Fehling's solution and 150 cc alcohol, together with a few fragments of porcelain, are placed in a 500 cc flask fitted with a doubly bored rubber stopper. In the one hole the tube containing the weighed substance is placed, whilst through the other passes the end of an inclined condenser. The contents of the flask are boiled and the open end of the condenser connected with a bent tube terminating in a short leg which dips below water. When no more air is expelled a measuring vessel full of water is placed over the end of the tube, and the vessel with the substance pressed into the flask by means of a rod. Continued boiling for a short time suffices to liberate all the nitrogen.

---

[1] H. Strache and S. Iritzer, M. **14**, 37. Holleman and de Vries, Rec. **10**, 229. De Vries, B. **27**, 1521; **28**, 2611. Petersen, Z. An. **5**, 2.

[2] H. Meyer, M. **18**, 404.

In some cases it is desirable to recover the acid on account of its rarity, or to remove it in order to facilitate the determination, as in the case of stearic acid, the potassium salt of which causes the liquid to froth over. This can be accomplished, if the acid is sparingly soluble in water or dilute hydrochloric acid, by boiling the hydrazide with concentrated hydrochloric acid, during several hours; the solution is made up to 100 cc, the organic acid removed by means of a dry filter, the first few drops of the filtrate rejected, and 50 cc of the remainder taken for the determination. This method of hydrolysis does not distinguish between hydrazides and hydrazones, as the latter are also acted upon by hydrochloric acid. Ortho- and paratolylhydrazides are oxidized in the same manner as phenylhydrazides, so that the method is also applicable to them.[1]

Platinic chloride oxidizes hydrazine hydrochloride in accordance with the equation:

$$N_2H_4 \cdot 2HCl + 2PtCl_4 \rightarrow N_2 + 6HCl + 2PtCl_2;$$

the evolved nitrogen is determined by the method described on p. 101.[2]

Hydrazine salts may be titrated by potassium permanganate in presence of sulphuric acid, provided the concentration of the latter is 6–12 per cent.[3] The reaction is represented by the equation

$$17N_2H_4 + 13O \rightarrow 13H_2O + 14NH_3 + 10N_2.$$

---

[1] M. **14**, 38.   [2] Curtius, J. pr. **147**, 37.
[3] Petersen, Z. An. **5**, 3.

(2) IODOMETRIC METHOD.[1]

Phenylhydrazine and iodine react in accordance with the following equation:

$$C_6H_5NH.NH_2 + 2I_2 \longrightarrow 3HI + N_2 + C_6H_5I.$$

The interaction is quantitative in highly dilute solution with iodine present in excess. The determination is made by adding to a known volume of N/10 iodine solution the highly dilute solution of the base or its hydrochloride, obtained by hydrolysis as described in the preceding section; the excess of iodine is then titrated in the ordinary manner.

In presence of dilute sulphuric acid iodic acid oxidizes phenylhydrazine and this reaction may also be employed for the determination. The strength of the iodic acid solution is ascertained by means of sulphurous acid of known titre; it is then added, in excess, to the highly dilute solution of phenylhydrazine and sulphuric acid, and the mixture again titrated.

To the above methods may be added the *titration* of phenylhydrazine with hydrochloric acid; methyl-orange is used as indicator, and tolerably accurate results are obtained.[2]

DETERMINATION OF THE NITRO-GROUP (NO₂).

(A) Titration Method.[3]

Organic nitro-compounds are reduced to amino-derivatives by the action of stannous chloride, in

---
[1] E. v. Meyer, J. pr. 149, 115.   [2] Strache and Iritzer.
[3] H. Limpricht, B. 11, 35.

presence of hydrochloric acid, in accordance with the equation

$$R.NO_2 + 3SnCl_2 + 6HCl \rightarrow R.NH_2 + 3SnCl_4 + 2H_2O;$$

the unchanged stannous chloride is determined by titration, and, from the quantity which has reacted, the number of nitro-groups in the original compound may be ascertained. Solution of iodine, or of potassium permanganate, is employed for the titration.[1]

Reagents Required.

(1) *Stannous Chloride Solution.* Tin (150 grams) is dissolved in concentrated hydrochloric acid, the clear liquid decanted, mixed with concentrated hydrochloric acid (50 cc), and diluted to 1 liter.

(2) *Sodium Carbonate Solution.* Anhydrous sodium carbonate (180 grams) and sodium potassium tartrate (240 grams) are dissolved in water and diluted to 1 liter.

(3) *Iodine Solution.* Iodine (12.54 grams) is dissolved in potassium iodide solution and the liquid made up to 1 liter, it will then be approximately $N/10$, if exactly so 1 cc = 0.0059 gram Sn = 0.000 655 gram NO_2.

(4) *Starch Solution.* This must be dilute, recently prepared, and filtered.

Potassium Permanganate Solution. It should be

---

[1] Jenssen, J. pr. **78**, 193. S. W. Young and R. E. Swain, J. Am. (1897), **19**, 812–814. Journ. Chem. Soc. (1898), **74**, ii, 186.

N/10, and may be used instead of the iodine, its strength being determined by means of iron.

(I) *Method of Determination for Non-volatile Compounds.*

After the titre of the stannous chloride has been ascertained, the nitro-compound (about 0.2 gram) is placed in a 100 cc glass-stoppered flask, stannous chloride solution (10 cc) added, and the liquid warmed during thirty minutes. When cool, the mixture is diluted to the mark, and, after shaking, 10 cc transferred to a beaker by means of a pipette; a little water is added, then the sodium carbonate solution, until the precipitate which first forms is wholly dissolved; after the addition of a little starch the iodine solution is run in until a permanent blue coloration is produced.

The results of the analysis are calculated according to the formula $NO_2 = (a - b).0.0007655$ gram, where $a =$ the number of cc of iodine solution equivalent to 1 cc of the stannous chloride solution, and $b =$ the number of cc of iodine solution required in the determination.

If it is desired to use the potassium permanganate 10 cc of the acid liquid, withdrawn as described above, is boiled with ferric chloride, and the ferrous chloride produced is determined in the ordinary manner.

(II) *Modified Method for Volatile Compounds.*

Volatile nitro-compounds are weighed in a test-tube about 30 cm by 8 mm, closed with a cork; the cork is removed, and the tube, together with the stannous chloride, placed in a second larger one, 20 cm by 13-15 mm, which is then sealed. The larger tube may be of thin-walled, readily fusible glass, as it will only be subjected to a very slight pressure. The tube is heated in the water-bath during 1-2 hours, and well shaken occasionally; it is then cooled, the contents completely washed into a 100-cc graduated flask, and treated in the manner described in the preceding section. The use of a sealed tube is sometimes advisable in the case of non-volatile compounds with which low results may be obtained by heating in the stoppered bottle.

(B) Diazo Method.[1]

Should the preceding method fail to give decisive results the nitro-compound must be reduced to the amino-derivative and this treated in the manner described on p. 87. As an example metanitrobenzaldehyde may be converted into metachlorobenzaldehyde at one operation.[2] It is dissolved in concentrated hydrochloric acid (6 parts), stannous chloride (4.5 parts) added, and after the reduction, without precipitating the tin, it is mixed with the calculated quantity of sodium nitrite and an equal weight of finely divided copper.

DETERMINATION OF THE IODOSO- (IO) AND IODOXY- (IO_2) GROUPS.

Iodoso- and iodoxy-compounds in presence of glacial acetic, of hydrochloric acid, or of dilute sulphuric acid liberate from potassium iodide an amount

---

[1] Gattermann, B. 23, 1222.   [2] Gattermann, *loc. cit.*

of iodine equivalent to their content of oxygen; one molecule of the former therefore liberates two, and of the latter four atoms of iodine. For the determination the substance is heated on the water-bath during four hours with acidified potassium iodide solution in a sealed tube from which the air has been expelled by carbonic anhydride.[1] The compound may also be digested on the water-bath with concentrated potassium iodide solution, glacial acetic acid, in fairly large quantity, and dilute sulphuric acid.[2] When the reaction is completed the liquid is titrated with N/10 sodium thiosulphate solution; no indicator is required. Whenever hydrochloric acid or sulphuric acid has been employed in the reduction, the iodide, which is produced, always retains some iodine in solution hence, during the titration, it is necessary to warm and shake the liquid until this has all been acted upon by the thiosulphate. The oxygen percentage content of the iodosy- and iodoxy-compounds is given by the formula $O = \dfrac{0.8 \cdot c \cdot 100}{1000\, s} = 0.08 \dfrac{c}{s}$, where s is the weight of the compound taken and c the number of cc of N/10 sodium thiosulphate employed.

DETERMINATION OF THE PEROXIDE GROUP $\left(\overset{\shortparallel}{C}\!\!<\!\!\overset{O}{\underset{O}{\cdot}}\right)$.

The oxygen of the acyl superoxides may be determined by means of stannous chloride in acid solution.

---

[1] V. Meyer and Wachter, B. **25**, 2632. P. Askenasy and V. Meyer, *Ibid.* **26**, 1355, *et seq.*

[2] Willgerodt, *Ibid.* **25**, 3495, *et seq.*

[3] Pechmann and Vanino, *Ibid.* **27**, 1512.

A known quantity of the peroxide is heated during about five minutes, in an atmosphere of carbonic anhydride, with a measured volume of a titrated, acidified stannous chloride solution. When the liquid is clear the remaining stannous chloride is determined by means of N/10 iodine solution.

THE IODINE NUMBER.[1]

This value expresses the quantity of iodine absorbed by one hundred parts of the substance, usually a fat or higher aliphatic acid. The acids of this series, such as oleic acid, ricinoleic acid, linoleic acid and linolenic acid, as well as their glycerides, absorb the first two, two, the others four and six atoms of iodine, bromine, or chlorine respectively, whilst the corresponding saturated compounds, under similar circumstances, are not affected. The reaction is carried out at the ordinary temperature, the substance being mixed with alcoholic iodine and mercuric chloride solutions.[2] The organic products are chloro-iodine additive compounds, some of which have been isolated and characterized.[3] The method is extensively employed in the technical investigation of fats, oils, waxes, resins, etheral oils, caoutchouc, etc., and is sometimes useful for scientific purposes, hence a brief description of the method of analysis is given here.

---

[1] Benedikt, "Analyse d. Fette und Wachsarten," III. Edition, p. 148. Allen, "Commercial Organic Analysis," vol. II, 3d Edition.

[2] Hübl, Dingl. 253, 281.

[3] R. Henriques and H. Künne. B. 32, 389.

Reagents.

(1) *Iodine Solution.* Iodine (25 grams) and mercuric chloride (30 grams) are each separately dissolved in 95 per cent alcohol (500 cc), free from fusel oil. The mercuric chloride solution is filtered if necessary, and the liquids mixed. The mixing must precede the use of the solution by 6–12 hours as, during this period, the titre rapidly changes.

(2) *Sodium Thiosulphate Solution.* The crystallized salt (24 grams) is dissolved in water and diluted to one liter. It is standardized in the following manner:[1] Potassium bichromate (3.8740 grams) is dissolved in water, diluted to one liter, and 20 cc of the liquid transferred to a stoppered bottle containing 10 cc of potassium iodide solution (10 per cent), and 5 cc hydrochloric acid; the liberated iodine is then titrated in the ordinary manner by means of sodium thiosulphate, starch being used as indicator; 1 cc of the above bichromate solution liberates 0.01 grams of iodine.

(3) *Chloroform.* Its purity is determined by a blank experiment.

(4) *Potassium Iodide Solution.* The salt is dissolved in ten parts of water.

(5) *Starch Solution.* This must be clear and recently prepared.

Method of Analysis.

The substance (0.15–1.0 gram) is mixed with chloroform (about 10 cc) in a 500–800 cc flask provided with

---

[1] Volhard.

a well-fitting glass-stopper. When the compound has dissolved the iodine solution (25 cc) is added by means of a pipette which must be manipulated so that equal quantities are delivered in each experiment. The flask is well shaken and more chloroform added if needful; should the liquid become almost colorless in a short time a second 25 cc of iodine solution is added, and this repeated, if necessary until, after the expiration of two hours, the liquid appears dark brown. The mixture is now allowed to remain during twelve hours at the ordinary temperature in the dark; it is then thoroughly mixed with at least 20 cc of potassium iodide solution and 300–500 cc of water, and titrated with the sodium thiosulphate solution, the liquid being constantly agitated; when only a faint color is visible in both the aqueous and chloroform solutions, starch is added and the titration completed. The production of a red precipitate of mercuric iodide, on the addition of water before the titration, indicates that too little potassium iodine has been employed, but this may be corrected by the immediate addition of more. A blank experiment must always be made with 25 cc of the iodine solution under exactly the same conditions as the test, and its titration must immediately precede or follow that of the actual determination.

Useful information is sometimes given by the *terebenthene number*.[1]

---

[1] J. Klimont, Ch. Ztg. (1894), No. 36, 37. Ch. R. (1894), 2, 2.

APPENDIX.

WEIGHT OF A CUBIC CENTIMETER OF HYDROGEN PERATURE OF 10°–25°.[1]

The observed height of the barometer is reduced to 0° by and 20°–25° respectively.

Height of barometer.	10° C.	11° C.	12° C.	13° C.	14° C.	15° C.	16° C.	17° C.
mm	mg	mg	mg	mg	mg	mg	mg	mg
700	0.07851	0.07816	0.07781	0.07746	0.07711	0.07675	0.07639	0.07603
702	0.07874	0.07839	0.07804	0.07769	0.07713	0.07697	0.07661	0.07625
704	0.07896	0.07861	0.07826	0.07791	0.07756	0.07720	0.07684	0.07647
706	0.07919	0.07884	0.07848	0.07813	0.07778	0.07742	0.07706	0.07670
708	0.07942	0.07907	0.07871	0.07836	0.07800	0.07774	0.07729	0.07692
710	0.07964	0.07929	0.07893	0.07858	0.07823	0.07787	0.07750	0.07714
712	0.07987	0.07952	0.07917	0.07881	0.07845	0.07809	0.07772	0.07736
714	0.08009	0.07975	0.07939	0.07903	0.07868	0.07832	0.07795	0.07759
716	0.08032	0.07997	0.07961	0.07924	0.07890	0.07854	0.07817	0.07781
718	0.08055	0.08019	0.07984	0.07948	0.07912	0.07876	0.07840	0.07803
720	0.08078	0.08043	0.08007	0.07971	0.07935	0.07899	0.07862	0.07825
722	0.08101	0.08065	0.08029	0.07993	0.07957	0.07921	0.07884	0.07847
724	0.08123	0.08087	0.08052	0.08016	0.07979	0.07943	0.07907	0.07869
726	0.08146	0.08110	0.08074	0.08038	0.08002	0.07965	0.07929	0.07891
728	0.08169	0.08133	0.08097	0.08061	0.08024	0.07987	0.07951	0.07913
730	0.08191	0.08156	0.08120	0.08083	0.08047	0.08010	0.07973	0.07936
732	0.08215	0.08179	0.08142	0.08106	0.08069	0.08032	0.07995	0.07958
734	0.08237	0.08201	0.08164	0.08129	0.08091	0.08055	0.08018	0.07980
736	0.08259	0.08224	0.08187	0.08151	0.08114	0.08077	0.08040	0.08002
738	0.08282	0.08246	0.08209	0.08173	0.08136	0.08099	0.08062	0.08024
740	0.08305	0.08269	0.08233	0.08196	0.08158	0.08122	0.08084	0.08047
742	0.08328	0.08291	0.08255	0.08218	0.08181	0.08144	0.08106	0.08069
744	0.08351	0.08314	0.08277	0.08240	0.08203	0.08166	0.08129	0.08091
746	0.08373	0.08337	0.08300	0.08263	0.08226	0.08189	0.08151	0.08113
748	0.08396	0.08360	0.08322	0.08285	0.08248	0.08211	0.08173	0.08135
750	0.08419	0.08382	0.08344	0.08308	0.08270	0.08234	0.08195	0.08158
752	0.08441	0.08404	0.08368	0.08331	0.08293	0.08256	0.08218	0.08180
754	0.08464	0.08428	0.08390	0.08353	0.08315	0.08278	0.08240	0.08202
756	0.08487	0.08450	0.08413	0.08376	0.08338	0.08301	0.08262	0.08224
758	0.08510	0.08472	0.08435	0.08398	0.08360	0.08323	0.08285	0.08246
760	0.08533	0.08496	0.08458	0.03420	0.08382	0.08345	0.08307	0.08269
762	0.08555	0.08518	0.08481	0.08443	0.08405	0.08367	0.08329	0.08291
764	0.08578	0.08541	0.08503	0.08465	0.08428	0.08389	0.08352	0.08313
766	0.08601	0.08563	0.08525	0.08487	0.08450	0.08412	0.08374	0.08335
768	0.08624	0.08586	0.08549	0.08511	0.08473	0.08434	0.08396	0.08357
770	0.08646	0.08608	0.08571	0.08533	0.08495	0.08466	0.08418	0.08380

[1] A. Baumann, Z. ang. Ch. *1891*, 210.

APPENDIX.

UNDER A PRESSURE OF 700-770 MM AND AT A TEMPERATURE

Value of $\dfrac{(b - \omega)0.089523}{760(1 + 0.00366t)}$.

subtracting 1, 2, or 3 mm for the temperatures $10°-12°$, $13°-19°$,

18° C.	19° C.	20° C.	21° C.	22° C.	23° C.	24° C.	25° C.	Height of barometer.
mg	mg	mg	mg	mg	mg	mg	mg	mm
0.07557	0.07529	0.07493	0.07455	0.07417	0.07380	0.07340	0.07300	700
0.07588	0.07552	0.07515	0.07477	0.07439	0.07401	0.07362	0.07322	702
0.07610	0.07574	0.07537	0.07499	0.07461	0.07422	0.07383	0.07344	704
0.07633	0.07595	0.07559	0.07521	0.07483	0.07444	0.07405	0.07366	706
0.07655	0.07618	0.07581	0.07543	0.07505	0.07466	0.07427	0.07387	708
0.07677	0.07640	0.07603	0.07565	0.07527	0.07487	0.07449	0.07409	710
0.07699	0.07662	0.07625	0.07587	0.07548	0.07509	0.07470	0.07431	712
0.07722	0.07684	0.07646	0.07608	0.07570	0.07531	0.07492	0.07452	714
0.07743	0.07706	0.07668	0.07630	0.07592	0.07553	0.07513	0.07473	716
0.07765	0.07728	0.07690	0.07652	0.07614	0.07574	0.07535	0.07495	718
0.07788	0.07749	0.07712	0.07674	0.07635	0.07596	0.07550	0.07516	720
0.07809	0.07772	0.07734	0.07696	0.07657	0.07618	0.07577	0.07538	722
0.07831	0.07794	0.07756	0.07718	0.07679	0.07640	0.07609	0.07560	724
0.07854	0.07816	0.07778	0.07740	0.07701	0.07661	0.07621	0.07582	726
0.07876	0.07838	0.07800	0.07762	0.07723	0.07683	0.07643	0.07604	728
0.07908	0.07860	0.07822	0.07784	0.07744	0.07705	0.07665	0.07624	730
0.07920	0.07882	0.07844	0.07805	0.07766	0.07727	0.07687	0.07646	732
0.07942	0.07904	0.07866	0.07827	0.07780	0.07748	0.07708	0.07668	734
0.07964	0.07926	0.07888	0.07849	0.07810	0.07770	0.07730	0.07689	736
0.07986	0.07948	0.07910	0.07871	0.07831	0.07792	0.07752	0.07711	738
0.08009	0.07970	0.07932	0.07893	0.07853	0.07813	0.07774	0.07732	740
0.08030	0.07992	0.07954	0.07915	0.07875	0.07835	0.07795	0.07754	742
0.08053	0.08014	0.07976	0.07937	0.07897	0.07857	0.07817	0.07776	744
0.08075	0.08036	0.07998	0.07959	0.07919	0.07879	0.07838	0.07797	746
0.08097	0.08058	0.08020	0.07981	0.07940	0.07900	0.07860	0.07819	748
0.08119	0.08080	0.08042	0.08002	0.07962	0.07922	0.07881	0.07840	750
0.08141	0.08102	0.08063	0.08024	0.07984	0.07944	0.07903	0.07862	752
0.08163	0.08124	0.08085	0.08046	0.08006	0.07966	0.07925	0.07883	754
0.08185	0.08146	0.08107	0.08068	0.08028	0.07987	0.07947	0.07905	756
0.08207	0.08168	0.08129	0.08090	0.08050	0.08009	0.07968	0.07927	758
0.08229	0.08190	0.08151	0.08112	0.08071	0.08031	0.07990	0.07949	760
0.08251	0.08212	0.08173	0.08134	0.08093	0.08052	0.08012	0.07970	762
0.08273	0.08234	0.08195	0.08155	0.08115	0.08074	0.08033	0.07992	764
0.08295	0.08256	0.08217	0.08177	0.08137	0.08096	0.08055	0.08013	766
0.08318	0.08278	0.08239	0.08199	0.08158	0.08118	0.08076	0.08034	768
0.08341	0.08301	0.08261	0.08221	0.08180	0.08139	0.08098	0.08056	770

TENSION OF AQUEOUS VAPOR.

0° C.	mm	0° C.	mm
10.0	9.165	18.0	15.357
10.5	9.474	18.5	15.845
11.0	9.792	19.0	16.346
11.5	10.120	19.5	16.861
12.0	10.457	20.0	17.391
12.5	10.804	20.5	17.935
13.0	11.162	21.0	18.495
13.5	11.530	21.5	19.069
14.0	11.908	22.0	19.659
14.5	12.298	22.5	20.265
15.0	12.699	23.0	20.888
15.5	13.112	23.5	21.528
16.0	13.536	24.0	22.184
16.5	13.972	24.5	22.858
17.0	14.421	25.0	23.550
17.5	14.882		

TABLE FOR THE VALUE OF $\dfrac{a}{1000-a}$. $a = 1 - a = 999$.[1]

	0	1	2	3	4	5	6	7	8	9
00	0.0000	010	020	030	040	050	060	071	081	091
01	101	111	122	132	142	152	163	173	183	194
02	204	215	225	235	246	256	267	278	288	299
03	309	320	331	341	352	363	373	384	395	406
04	417	428	438	449	460	471	482	493	504	515
05	526	537	549	560	571	582	593	605	616	627
06	638	650	661	672	684	695	707	718	730	741
07	753	764	776	788	799	811	823	834	846	858
08	0.0870	881	893	905	917	929	941	953	965	977
09	989	*001	*013	*025	*038	*050	*062	*074	*087	*099
10	0.1111	124	136	148	151	173	186	198	211	223
11	236	249	261	274	287	299	312	325	338	351
12	364	377	390	403	416	429	442	455	468	481
13	494	508	528	534	547	564	574	588	601	614
14	628	641	655	669	682	696	710	723	737	751
15	765	779	793	806	820	834	848	862	877	891
16	905	919	933	947	962	976	990	*005	*019	*034
17	0.2048	083	077	092	107	121	136	151	166	180
18	195	210	225	240	255	270	285	300	315	331
19	346	361	376	392	407	422	438	453	469	484

[1] Obach—Ostwald, Z. **II**. 566.

APPENDIX. 119

TABLE FOR THE VALUE OF $\dfrac{a}{1000-a}$. (*Continued.*)

	0	1	2	3	4	5	6	7	8	9
20	0.2500	516	531	547	563	579	595	610	626	642
21	658	674	690	707	723	739	755	771	788	804
22	821	837	854	870	887	903	920	937	953	970
23	987	*004	*021	*038	*055	*072	*089	*106	*123	*141
24	0.3158	175	193	210	228	245	263	280	298	316
25	333	351	369	387	405	423	441	459	477	495
26	514	532	550	569	587	605	624	643	661	680
27	699	717	736	755	774	793	812	831	850	870
28	889	908	928	947	967	986	*006	*025	*045	*065
29	0.4085	104	124	144	164	184	205	225	245	265
30	286	306	327	347	365	389	409	430	451	472
31	493	514	535	556	577	599	620	641	663	684
32	706	728	749	771	793	815	837	859	881	903
33	925	948	970	993	*015	*038	*060	*083	*106	*129
34	0.5152	175	198	221	244	267	291	314	337	361
35	385	408	432	456	480	504	528	552	576	601
36	625	650	674	699	721	748	773	798	813	848
37	873	898	924	949	974	*000	*026	*051	*077	*163
38	0.6129	155	181	208	234	260	287	313	340	367
39	393	420	447	475	502	529	556	584	611	639
40	667	695	722	750	779	807	835	863	892	921
41	949	978	*007	*036	*065	*094	*123	*153	*182	*212
42	0.7241	271	301	331	361	391	422	452	483	513
43	544	575	606	637	668	699	731	762	*794	825
44	857	889	921	953	986	*018	*051	*083	116	*149
45	0.8182	215	248	282	315	349	382	416	450	484
46	519	553	587	622	657	692	727	762	797	832
47	868	904	939	975	*011	*048	*084	*121	*157	*194
48	0.9231	268	305	342	380	418	455	493	531	570
49	608	646	685	724	763	802	841	881	920	960
50	1.000	004	008	012	016	020	024	028	033	037
51	041	045	049	053	058	062	066	070	075	079
52	083	088	092	096	101	105	110	114	119	123
53	128	132	137	141	146	151	155	160	165	169
54	174	179	183	188	193	198	203	208	212	217
55	222	227	232	237	242	247	252	257	262	268
56	273	278	283	288	294	299	304	309	315	320
57	326	331	336	342	347	353	358	364	370	375
58	381	387	392	398	404	410	415	421	427	433
59	439	445	451	457	463	469	475	484	488	494

APPENDIX.

TABLE FOR THE VALUE OF $\dfrac{a}{1000-a}$. (*Continued.*)

	0	1	2	3	4	5	6	7	8	9
60	1.500	506	513	519	525	532	538	545	551	556
61	564	571	577	584	591	597	604	611	618	625
62	632	639	646	653	660	667	674	681	688	695
63	703	710	717	725	732	740	747	755	762	770
64	778	786	793	801	809	817	825	833	841	849
65	857	865	874	882	890	899	907	915	924	933
66	941	950	959	967	976	985	994	*003	*012	*021
67	2.030	040	049	058	067	077	086	096	106	115
68	125	135	145	155	165	175	185	195	205	215
69	226	236	247	257	268	279	289	300	311	322
70	333	344	356	367	378	390	401	413	425	436
71	448	460	472	484	497	509	521	534	546	559
72	571	584	597	610	623	636	650	663	676	690
73	704	717	731	745	759	774	788	802	817	831
74	846	861	876	891	906	922	937	953	968	984
75	3.000	016	032	049	065	082	098	115	132	149
76	167	184	202	219	237	255	274	292	310	329
77	348	367	386	405	425	444	464	484	505	525
78	545	566	587	608	630	651	673	695	717	739
79	762	785	808	831	854	878	902	926	950	975
80	4.000	025	051	076	102	128	155	181	208	236
81	263	291	319	348	376	405	435	465	495	525
82	556	587	618	650	682	714	747	780	814	848
83	882	917	952	988	*024	*061	*098	*135	*173	*211
84	5.250	289	329	369	410	452	494	536	579	623
85	667	711	757	803	849	897	944	993	*042	*092
86	6.143	194	246	299	353	407	463	519	576	654
87	692	752	813	874	937	*000	*065	*130	*197	*264
88	7.333	403	475	547	621	696	772	850	929	*009
89	8.091	174	259	346	434	524	615	769	804	901
90	9.000	101	204	309	417	526	638	753	870	989
91	10.11	10.33	10.36	10.49	10.63	10.77	10.90	11.05	11.20	11.35
92	11.50	11.66	11.82	11.99	12.16	12.33	12.51	12.70	12.89	13 08
93	13.29	13.49	13.71	13.93	14.15	14.38	14.63	14.87	15.13	15.39
94	15.67	15.95	16.24	16.54	16.86	17.18	17.52	17.87	18.23	18.61
95	19.00	19.41	19.83	20.28	20.74	21.22	21.73	22.26	22.81	23.39
96	24.00	24.64	25.32	26.03	26.78	27.57	28.41	29.30	30.25	31.26
97	32.33	33.48	34.71	36.04	37.46	39.00	40.67	42.48	44.45	46.62
98	49.00	51.6	54.6	57.8	61.5	65.7	70.4	75.9	82.3	89.9
99	99.0	110	124	142	166	199	249	332	499	999

INDEX OF AUTHORS.

A

Allen, A. H. 111
Anderlini 4, 61
Angeli, A. 88
Askenasy, P. 110
Auwers, K. 72

B

Baeyer, A. v. 74, 77, 78
Bamberger, E. 19, 62, 64, 72, 103
Bamberger, M. 35, 37
Barth 12, 20
Barus 46
Baum 62, 74
Baumann 18, 57, 59, 116
Beckmann 8, 32, 40
Benedikt 4, 10, 35, 36, 37, 65, 68, 111
Berthelot, D. 50
Biginelli 74
Blau, Fr. 82
Boeris 99
Bouveault 91
Buchka 4, 15

C

INDEX OF AUTHORS.

Cain.. 90
Cavazzi... 88
Ciamician... 14, 99
Claisen, L.................................. 6, 20, 22, 23
Claus.. 74, 91
Cohen, E.. 47
Collie, N... 89
Crismer... 73
Curtius.............................. 86, 100, 101, 102, 105

D

Danckworth.. 10, 18
Davies.. 74
Dedichen.. 86
Delépine.. 83
Deninger, A... 6, 21
Diamant, J... 8
Dobriner, P... 17

E

Ebert... 49
Eckhardt.. 42
Ehmann, L... 35
Elbers.. 62
Ephraim... 62
Erb.. 90, 91
Erdmann.. 11, 15, 30
Erk... 15
Erlich... 4

F

Feist.. 6, 21
Feit.. 74
Feitler... 88
Fischer, E.............................. 45, 60, 63, 64, 93
Franchimont.. 8
Fresenius... 15
Freund.. 45

INDEX OF AUTHORS.

Frobenius 103
Fuchs, F. 52, 56

G

Garelli 73
Gattermann 29, 87, 88, 109
Ghiro 4
Giraud, H. 93
Goldschmidt, H. 103
Goldschmiedt 12, 14, 17, 20, 21, 31, 51
Graebe 4, 28
Grandmougin 86
Green, A. G. 85
Gregor, J. 39
Griess, P. 86
Gröger, M. 59
Grüssner 35
Gumpert 32
Guyot 28

H

Hagen 44, 45
Haitinger 43
Haller 28
Hautzsch 91, 103
Harpe, De la 84, 93
Harries, C. 74
Hawkins 86
Hemmelmayr 14, 17, 21, 51, 62
Henriques, R. 111
Herzfeld 13, 80
Herzig, J. 5, 11, 14, 15, 28, 38, 73, 94, 99
Heuser 75
Heyl, G. 92
Hinsberg 24, 27
Hirsch, R. 85
Hoffmann, C. 74
Hoffmann, E. 20
Hofmann, A. W. 31, 89, 90

Hölle.. 62
Hollemann.. 104
Homolka... 42, 71
Hörmann, O.. 7
Hübl... 111
Huth... 30
Hyde, E.. 64

I

Iritzer, S... 104, 106

J

Jackson, F. L.. 23
Jacobson, P............................... 29, 80, 90, 92
Janny.. 70
Jassoy... 27
Jeaneraud.. 74
Jehn, C.. 52
Jenssen.. 107
Jones, H... 46
Just... 62

K

Kehrmann... 72, 73
Kinnicutt, L. P.. 86
Klimont, J... 113
Klobukowsky... 8, 11
Knoevenagel.. 103
Knop... 57
Knorr.. 26
Kohlrausch... 47
Kormann, W... 81
Kostanecki... 28, 73
Kraus.. 64
Krüger... 64, 77
Künne, H... 111
Kux.. 57

INDEX OF AUTHORS. 125

L

La Coste	8
Landsiedl	5
Lassar-Cohn	42
Lehmann	74
Lieben	10, 12, 43
Liebermann, C	7, 14, 21, 44, 45
Limpricht, H	106
Lossen	18, 19
Lucas	91

M

McIlhiney, P. C.	52
Marchlewsky	28
Meldola	86
Menschutkin	83
Merz, A.	103
Meyer, E. v.	106
Meyer, H	16, 25, 38, 57, 94, 99, 104
Meyer, R	16, 25, 62
Meyer, V	20, 29, 44, 62, 70, 74, 80, 90, 91, 92, 110
Meyer	42
Michael, H. A.	8, 16
Michaelis	63
Michel, O.	86
Munch	91
Münchmeyer	62

N

Nef, J. U.	63, 74, 86
Neufeld, A.	65
Nietzki	72
Nölting	86

O

Obach	118
Ostwald, W.	46, 48, 118

Otto... 23
Overton, B.. 61, 63
Overton, R.. 65

P

Panormow... 18
Patterson... 50
Pechmann, v........................... 18, 28, 62, 103, 110
Perkin, W. H., Sen..................................... 28
Perkin, W. H., Jun..................................... 42
Petersen... 104, 105
Petraczek... 71
Pomeranz.. 38
Pum, G... 24, 37

R

Radziszewski.. 91
Raschig... 73
Regnault... 103
Reverdin.. 84, 93
Richards, T. W.. 44
Rideal, S... 85
Rolfe, G. W... 23

S

Sachsse... 81
Sandmeyer... 87
Sarauw... 4
Saul, E... 62
Schall.. 15
Schiaparelli, C... 24
Schiff... 5, 12, 14
Schlömann.. 22, 24
Schmidt, G.. 29
Schmiedeberg.. 42
Schmolge.. 13
Schöpf.. 22
Schotten.. 22, 23, 24

INDEX OF AUTHORS.

Schreder.. 20
Schultz... 11
Schulze... 15
Schunk.. 28
Schützenberger.. 13
Seelig.. 5, 62, 72, 91
Silber.. 14
Sisley, P... 16
Skraup.. 19
Smith, Alex... 42
Smoeger... 13
Snape... 31, 32
Speier.. 45
Stange.. 75
Stallburg... 90
Strache, H.. 65, 68, 104, 106
Strouhal.. 46
Sudborough.. 90, 91, 92
Swain, R. E.. 107

T

Tafel... 63
Tessmer.. 31, 32
Thiele.. 74, 75, 78, 79
Thompson.. 22, 23
Thoms... 62
Thorp... 72
Tickle, T... 89
Tiemann, F.. 64, 71, 77
Tingle, A.. 45, 63
Tingle, J. Bishop................................ 45, 63, 74

U

Ulzer... 10

V

Vanin 110
Vohl 52
Volhard 40, 71, 112
Vongerichten 26, 32
Vortmann 11
Vries, de 104

W

Wachter 110
Wagner 57
Wallbaum 91
Wiedemann 49
Willgerodt 110
Wislicenus 6, 14
Wohl 71
Wolff 80
Wright 10

Y

Young, S. W 107

Z

Zanoli 31
Zeisel, S 10, 12, 28, 33, 37, 38, 41, 73

INDEX OF SUBJECTS.

A

Acetic acid	3
glacial	5, 8
anhydride	5, 7
Acetylation, methods of	5
Acetyl chloride	5, 6
derivatives, isolation of	9
preparation of	5
groups, determination of	9
(additive method)	14
(distillation method)	15
(potassium acetate method)	14
Acids, electrolytic conductivity of	46
etherification of	44
titration of	43
Acylation	3
Aliphatic amino groups, determination of	81
diazo-compounds	100
Alkylation	3
of hydroxyl groups	28
Amidodimethylaniline (p) derivatives	80
group, determination of	91
guanidine derivatives, preparation of	78
picrate derivatives, preparation of	80
salts, preparation of	78
Amines, acetylation of	83, 89
salts of	82, 89
Amino groups, determination of	81
Aqueous vapor, tension of	118

INDEX OF SUBJECTS.

Aromatic amino groups, determination of.................. 83
        diazo-compounds............................... 103
Authors, index of....................................... 121
Azoinide method, for determination of amino group......... 86

B

Barium hydroxide, hydrolysis by....................... 9, 11
Basicity of acids, determination of by ammonia method....... 52
                                    carbonate method.......... 51
                                    hydrogen sulphide method   52
                                    iodine-oxygen method..... 57
Benzene and water, tension of.............................. 68
Benzoic acid... 3
        acids, substituted.................................. 3
        anhydride..................................... 17, 21
Benzoyl chloride.. 17, 18
        derivatives, analysis of........................... 24
                preparation of....................... 17
Benzyl derivatives.. 28
p-Brombenzoic anhydride......................... 17, 22, 23
o-Brombenzoyl chloride............................. 17, 22, 23
p-Brombenzoyl chloride............................. 17, 22, 23

C

Carbamates.. 3
        preparation of................................. 29
Carbamyl chloride, preparation of........................ 29
Carbonyl, determination of............................... 60
Carboxyl, determination of............................... 41
                  indirect............................... 65
Chloracetyl chloride.................................... 5, 8
1 : 2 : 4-Chlorodinitrobenzene........................... 32
Cyanide group, determination of......................... 90

D

Diazo-compounds, aliphatic.............................. 100
                aromatic.............................. 103
                preparation of....................... 84
        group, determination of........................... 100

Diazonium derivatives.. 103
Diazomethane as reagent for hydroxyl....................... 28
Diphenylcarbamyl chloride, preparation of................... 30

E

Electrolytic conductivity of sodium salts..................... 46
Etherification of acids.. 44
Ethoxyl, determination of....................................... 41
    and methoxyl, differentiation of.................... 40
Ethylimide and methylimide, differentiation of............. 99
Ethylimide, determination of................................... 99

G

Gattermann-Sandmeyer's reaction.............................. 87

H

Hydrazide group, determination of............................ 103
Hydrazides, oxidation of.. 104
Hydrazones, substituted, preparation of.............. 63, 64
Hydrochloric acid, hydrolysis by........................... 10, 13
Hydrogen, weight of a cc of.................................... 116
Hydriodic acid, hydrolysis by.............................. 10, 14
Hydrolytic methods for determination of acetyl groups...... 9
Hydroxyl, determination of..................................... 3

I

Imide group, determination of.................................. 92
Imides, acetylation of... 92
    alkylation of.. 93
    salts of... 94
Index of authors... 121
    subjects... 129
Introduction.. 1
Iodine number.. 111
Iodoso-group, determination of................................ 109
Iodoxy-group, determination of................................ 109
Isobutyric acid... 3
    anhydride.. 27
Isobutyryl derivatives... 27

M

Magnesia, hydrolysis by 10, 12
Metanitrobenzoyl chloride 17, 22, 23
Methoxyl, determination of (Zeisel's method)............... 33
                                   modified....... 39
       and ethoxyl, differentiation of.................. 40
Methylimide and ethylimide, differentiation of............. 99
                          determination of............... 94
      determination of in presence of methoxyl...... 97

N

Nitrile group, determination of............................ 90
m-Nitrobenzoyl chloride............................ 17, 22, 23
Nitro-group, determination of............................. 106

O

Orthobromobenzoyl chloride......................... 17, 22, 23
Oximes, preparation of.................................... 70

P

Paramidodimethylaniline derivatives....................... 80
Parabrombenzoic anhydride......................... 17, 22, 23
Parabrombenzoyl chloride.......................... 17, 22, 23
Parabromphenylhydrazine, preparation of.................. 63
Peroxide group, determination of.......................... 110
Phenylacetic acid... 3
Phenylacetyl chloride..................................... 27
Phenylcarbamic acid derivatives, preparation of........... 31
Phenylcarbamates... 3
Phenylhydrazones, preparation of.......................... 60
              substituted, preparation of............ 63, 64
Phenylisocyanate, action of, on hydroxyl................. 31
              preparation of......................... 31
Phenylsulphonic acid...................................... 3
            chloride........................... 17, 23, 24
Phosphoric acid as reagent................................ 15
Phosphoric acid derivatives............................... 27

Potassium hydroxide, hydrolysis by........................ 9, 10
    hydroxylamine sulphonate as reagent............ 73
Propionic acid... 3
    anhydride... 27
Propionyl derivatives....................................... 27

S

Salts of acids, analysis of.................................. 42
    bases, preparation of............................. 82, 83
Sandmeyer-Gattermann's reaction........................... 87
Semicarbazide salts, preparation of.......................... 75
Semicarbazones, preparation of........................... 74, 77
Sodium acetate.. 5, 7
    benzoate.. 17, 21
    hydroxide, hydrolysis by......................... 9, 10
Stearic anhydride... 27
Substituted benzoic acids............................. 17, 22, 23
    acylation by means of.............. 23
    hydrazones, preparation of.................. 63, 64
    phenylhydrazones, preparation of........... 63, 64
Sulphuric acid, hydrolysis by.............................. 10, 13

T

Tables.. 116 *et seq.*
Table of tension of benzene and water...................... 68
    water... 118
    weight of a cc of hydrogen.......................... 116

W

Water, hydrolysis by 9, 10
Water and benzene, table of tension of..................... 68
    hydrolysis by.................................... 9, 10
    table of tension of............................. 118

SHORT-TITLE CATALOGUE

OF THE

PUBLICATIONS

OF

JOHN WILEY & SONS,

NEW YORK.

LONDON: CHAPMAN & HALL, LIMITED.

ARRANGED UNDER SUBJECTS.

Descriptive circulars sent on application.
Books marked with an asterisk are sold at *net* prices only.
All books are bound in cloth unless otherwise stated.

AGRICULTURE.

CATTLE FEEDING—DAIRY PRACTICE—DISEASES OF ANIMALS—GARDENING, ETC.

Armsby's Manual of Cattle Feeding.................12mo,	$1	75
Downing's Fruit and Fruit Trees......................8vo,	5	00
Grotenfelt's The Principles of Modern Dairy Practice. (Woll.) 12mo,	2	00
Kemp's Landscape Gardening........................12mo,	2	50
Loudon's Gardening for Ladies. (Downing.).........12mo,	1	50
Maynard's Landscape Gardening....................12mo,	1	50
Steel's Treatise on the Diseases of the Dog...........8vo,	3	50
" Treatise on the Diseases of the Ox.............8vo,	6	00
Stockbridge's Rocks and Soils........................8vo,	2	50
Woll's Handbook for Farmers and Dairymen.........12mo,	1	50

ARCHITECTURE.

BUILDING—CARPENTRY—STAIRS—VENTILATION—LAW, ETC.

Berg's Buildings and Structures of American Railroads.....4to,	7	50
Birkmire's American Theatres—Planning and Construction.8vo,	3	00
" Architectural Iron and Steel..................8vo,	3	50
" Compound Riveted Girders...................8vo,	2	00
" Skeleton Construction in Buildings...........8vo,	3	00

1

Birkmire's Planning and Construction of High Office Buildings. 8vo,	$3	50
Carpenter's Heating and Ventilating of Buildings..........8vo,	3	00
Freitag's Architectural Engineering................8vo,	2	50
Gerhard's Sanitary House Inspection..................16mo,	1	00
" Theatre Fires and Panics.....................12mo,	1	50
Hatfield's American House Carpenter...................8vo,	5	00
Holly's Carpenter and Joiner................................18mo,		75
Kidder's Architect and Builder's Pocket-book...16mo, morocco,	4	00
Merrill's Stones for Building and Decoration.............8vo,	5	00
Monckton's Stair Building—Wood, Iron, and Stone........4to,	4	00
Wait's Engineering and Architectural Jurisprudence.......8vo,	6	00
Sheep,	6	50
Worcester's Small Hospitals—Establishment and Maintenance, including Atkinson's Suggestions for Hospital Architecture...12mo,	1	25
World's Columbian Exposition of 1893.............Large 4to,	2	50

ARMY, NAVY, Etc.
MILITARY ENGINEERING—ORDNANCE—LAW, ETC.

Bourne's Screw Propellers.............................4to,	5	00
* Bruff's Ordnance and Gunnery........................8vo,	6	00
Chase's Screw Propellers...........................8vo,	3	00
Cooke's Naval Ordnance 8vo,	12	50
Cronkhite's Gunnery for Non-com. Officers.....32mo, morocco,	2	00
* Davis's Treatise on Military Law......................8vo,	7	00
Sheep,	7	50
* " Elements of Law..............................8vo,	2	50
De Brack's Cavalry Outpost Duties. (Carr.)....32mo, morocco,	2	00
Dietz's Soldier's First Aid..................16mo, morocco,	1	25
* Dredge's Modern French Artillery....Large 4to, half morocco,	15	00
" Record of the Transportation Exhibits Building, World's Columbian Exposition of 1893..4to, half morocco,	10	00
Durand's Resistance and Propulsion of Ships..............8vo,	5	00
Dyer's Light Artillery................................12mo,	3	00
Hoff's Naval Tactics................................8vo,	1	50
* Ingalls's Ballistic Tables.............................8vo,	1	50
" Handbook of Problems in Direct Fire..............8vo,	4	00

Mahan's Permanent Fortifications. (Mercur.).8vo, half morocco,	$7 50
Mercur's Attack of Fortified Places....................12mo,	2 00
" Elements of the Art of War.....................8vo,	4 00
Metcalfe's Ordnance and Gunnery..........12mo, with Atlas,	5 00
Murray's A Manual for Courts-Martial........16mo, morocco,	1 50
" Infantry Drill Regulations adapted to the Springfield Rifle, Caliber .45.....................32mo, paper,	10
*Phelps's Practical Marine Surveying.....................8vo,	2 50
Powell's Army Officer's Examiner.....................12mo,	4 00
Sharpe's Subsisting Armies..................32mo, morocco,	1 50
Very's Navies of the World...............8vo, half morocco,	3 50
Wheeler's Siege Operations...............................8vo,	2 00
Winthrop's Abridgment of Military Law................12mo,	2 50
Woodhull's Notes on Military Hygiene..................16mo,	1 50
Young's Simple Elements of Navigation.......16mo, morocco,	2 00
" " " " " first edition........	1 00

ASSAYING.

SMELTING—ORE DRESSING—ALLOYS, ETC.

Fletcher's Quant. Assaying with the Blowpipe..16mo, morocco,	1 50
Furman's Practical Assaying............................8vo,	3 00
Kunhardt's Ore Dressing................................8vo,	1 50
O'Driscoll's Treatment of Gold Ores.....................8vo,	2 00
Ricketts and Miller's Notes on Assaying.................8vo,	3 00
Thurston's Alloys, Brasses, and Bronzes................8vo,	2 50
Wilson's Cyanide Processes............................12mo,	1 50
" The Chlorination Process.....................12mo,	1 50

ASTRONOMY.

PRACTICAL, THEORETICAL, AND DESCRIPTIVE.

Craig's Azimuth...4to,	3 50
Doolittle's Practical Astronomy..........................8vo,	4 00
Gore's Elements of Geodesy.............................8vo,	2 50
Hayford's Text-book of Geodetic Astronomy.............8vo,	3 00
* Michie and Harlow's Practical Astronomy..............8vo,	3 00
* White's Theoretical and Descriptive Astronomy........12mo,	2 00

BOTANY.

GARDENING FOR LADIES, ETC.

Baldwin's Orchids of New England..............Small 8vo,	$1	50
Loudon's Gardening for Ladies. (Downing.)............12mo,	1	50
Thomé's Structural Botany................................16mo,	2	25
Westermaier's General Botany. (Schneider.)............8vo,	2	00

BRIDGES, ROOFS, Etc.

CANTILEVER—DRAW—HIGHWAY—SUSPENSION.

(See also ENGINEERING, p. 7.)

Boller's Highway Bridges...8vo,	2	00
* " The Thames River Bridge................4to, paper,	5	00
Burr's Stresses in Bridges..8vo,	3	50
Crehore's Mechanics of the Girder........................8vo,	5	00
Dredge's Thames Bridges............. ....7 parts, per part,	1	25
Du Bois's Stresses in Framed Structures............Small 4to,	10	00
Foster's Wooden Trestle Bridges..........................4to,	5	00
Greene's Arches in Wood, etc..............................8vo,	2	50
" Bridge Trusses................................8vo,	2	50
" Roof Trusses..................................8vo,	1	25
Howe's Treatise on Arches 8vo,	4	00
Johnson's Modern Framed Structures..............Small 4to,	10	00
Merriman & Jacoby's Text-book of Roofs and Bridges. Part I., Stresses..... ....8vo,	2	50
Merriman & Jacoby's Text-book of Roofs and Bridges. Part II., Graphic Statics 8vo,	2	50
Merriman & Jacoby's Text-book of Roofs and Bridges. Part III., Bridge Design........................8vo,	2	50
Merriman & Jacoby's Text-book of Roofs and Bridges. Part IV., Continuous, Draw, Cantilever, Suspension, and Arched Bridges..........................8vo,	2	50
* Morison's The Memphis Bridge................Oblong 4to,	10	00
Waddell's Iron Highway Bridges.... ....8vo,	4	00
" De Pontibus (a Pocket-book for Bridge Engineers). 16mo, morocco,	3	00
Wood's Construction of Bridges and Roofs..............8vo,	2	00
Wright's Designing of Draw Spans. Parts I. and II..8vo, each	2	50
" " " " " Complete............8vo,	3	50

CHEMISTRY.
Qualitative—Quantitative—Organic—Inorganic, Etc.

Adriance's Laboratory Calculations............12mo,	$1	25
Allen's Tables for Iron Analysis....................8vo,	3	00
Austen's Notes for Chemical Students............12mo,	1	50
Bolton's Student's Guide in Quantitative Analysis.......8vo,	1	50
Classen's Analysis by Electrolysis. (Herrick and Boltwood.).8vo,	3	00
Crafts's Qualitative Analysis. (Schaeffer.)............12mo,	1	50
Drechsel's Chemical Reactions. (Merrill.)............12mo,	1	25
Fresenius's Quantitative Chemical Analysis. (Allen.).......8vo,	6	00
" Qualitative " " (Johnson.).....8vo,	3	00
" " " " (Wells.) Trans. 16th German Edition............................8vo,	5	00
Fuertes's Water and Public Health..................12mo,	1	50
Gill's Gas and Fuel Analysis......................12mo,	1	25
Hammarsten's Physiological Chemistry. (Mandel.)........8vo,	4	00
Helm's Principles of Mathematical Chemistry. (Morgan).12mo,	1	50
Kolbe's Inorganic Chemistry......................12mo,	1	50
Ladd's Quantitative Chemical Analysis..............12mo,	1	00
Landauer's Spectrum Analysis. (Tingle.)..............8vo,	3	00
Löb's Electrolysis and Electrosynthesis of Organic Compounds. (Lorenz.)..12mo,	1	00
Mandel's Bio-chemical Laboratory..................12mo,	1	50
Mason's Water-supply..........................8vo,	5	00
" Examination of Water.....................12mo,	1	25
Meyer's Organic Analysis. (Tingle.) (*In the press.*)		
Miller's Chemical Physics.........................8vo,	2	00
Mixter's Elementary Text-book of Chemistry...........12mo,	1	50
Morgan's The Theory of Solutions and its Results.......12mo,	1	00
" Elements of Physical Chemistry...........12mo,	2	00
Nichols's Water-supply (Chemical and Sanitary)..........8vo,	2	50
O'Brine's Laboratory Guide to Chemical Analysis.........8vo,	2	00
Perkins's Qualitative Analysis.....................12mo,	1	00
Pinner's Organic Chemistry. (Austen.)..............12mo,	1	50
Poole's Calorific Power of Fuels.....................8vo,	3	00
Ricketts and Russell's Notes on Inorganic Chemistry (Non-metallic).................... Oblong 8vo, morocco,		75
Ruddiman's Incompatibilities in Prescriptions...........8vo,	2	00

Schimpf's Volumetric Analysis..........12mo,	$2	50
Spencer's Sugar Manufacturer's Handbook.....16mo, morocco,	2	00
" Handbook for Chemists of Beet Sugar Houses. 16mo, morocco,	3	00
Stockbridge's Rocks and Soils..........8vo,	2	50
Tillman's Descriptive General Chemistry. (*In the press.*)		
Van Deventer's Physical Chemistry for Beginners. (Boltwood.) 12mo,	1	50
Wells's Inorganic Qualitative Analysis..........12mo,	1	50
" Laboratory Guide in Qualitative Chemical Analysis. 8vo,	1	50
Whipple's Microscopy of Drinking-water..........8vo,	3	50
Wiechmann's Chemical Lecture Notes..........12mo,	3	00
" Sugar Analysis..........Small 8vo,	2	50
Wulling's Inorganic Phar. and Med. Chemistry..........12mo,	2	00

DRAWING.

ELEMENTARY—GEOMETRICAL—MECHANICAL—TOPOGRAPHICAL.

Hill's Shades and Shadows and Perspective..........8vo,	2	00
MacCord's Descriptive Geometry..........8vo,	3	00
" Kinematics..........8vo,	5	00
" Mechanical Drawing..........8vo,	4	00
Mahan's Industrial Drawing. (Thompson.)........2 vols., 8vo,	3	50
Reed's Topographical Drawing. (H. A.)..........4to,	5	00
Reid's A Course in Mechanical Drawing..........8vo,	2	00
" Mechanical Drawing and Elementary Machine Design. 8vo. (*In the press.*)		
Smith's Topographical Drawing. (Macmillan.)..........8vo,	2	50
Warren's Descriptive Geometry..........2 vols., 8vo,	3	50
" Drafting Instruments..........12mo,	1	25
" Free-hand Drawing..........12mo,	1	00
" Linear Perspective..........12mo,	1	00
" Machine Construction..........2 vols., 8vo,	7	50
" Plane Problems..........12mo,	1	25
" Primary Geometry..........12mo,		75
" Problems and Theorems..........8vo,	2	50
" Projection Drawing..........12mo,	1	50

Warren's Shades and Shadows............8vo,	$3 00	
" Stereotomy—Stone-cutting............8vo,	2 50	
Whelpley's Letter Engraving............12mo,	2 00	

ELECTRICITY AND MAGNETISM.

ILLUMINATION—BATTERIES—PHYSICS—RAILWAYS.

Anthony and Brackett's Text-book of Physics. (Magic.) Small 8vo,	3 00
Anthony's Theory of Electrical Measurements............12mo,	1 00
Barker's Deep-sea Soundings............8vo,	2 00
Benjamin's Voltaic Cell............8vo,	3 00
" History of Electricity............8vo,	3 00
Classen's Analysis by Electrolysis. (Herrick and Boltwood.) 8vo,	3 00
Cosmic Law of Thermal Repulsion............12mo,	75
Crehore and Squier's Experiments with a New Polarizing Photo-Chronograph............8vo,	3 00
Dawson's Electric Railways and Tramways. Small, 4to, half morocco,	12 50
*Dredge's Electric Illuminations....2 vols., 4to, half morocco,	25 00
" " " Vol. II............4to,	7 50
Gilbert's De magnete. (Mottelay.)............8vo,	2 50
Holman's Precision of Measurements............8vo,	2 00
" Telescope-mirror-scale Method............Large 8vo,	75
Löb's Electrolysis and Electrosynthesis of Organic Compounds. (Lorenz.)............12mo,	1 00
*Michie's Wave Motion Relating to Sound and Light,......8vo,	4 00
Morgan's The Theory of Solutions and its Results............12mo,	1 00
Niaudet's Electric Batteries. (Fishback.)............12mo,	2 50
Pratt and Alden's Street-railway Road-beds............8vo,	2 00
Reagan's Steam and Electric Locomotives............12mo,	2 00
Thurston's Stationary Steam Engines for Electric Lighting Purposes............8vo,	2 50
*Tillman's Heat............8vo,	1 50

ENGINEERING.

Civil—Mechanical—Sanitary, Etc.

(*See also* Bridges, p. 4; Hydraulics, p. 9; Materials of Engineering, p. 10; Mechanics and Machinery, p. 12; Steam Engines and Boilers, p. 14.)

Baker's Masonry Construction..........................8vo,	$5	00
" Surveying Instruments......................12mo,	3	00
Black's U. S. Public Works....................Oblong 4to,	5	00
Brooks's Street-railway Location..............16mo, morocco,	1	50
Butts's Civil Engineers' Field Book............16mo, morocco,	2	50
Byrne's Highway Construction...........................8vo,	5	00
" Inspection of Materials and Workmanship.......16mo,	3	00
Carpenter's Experimental Engineering 8vo,	6	00
Church's Mechanics of Engineering—Solids and Fluids....8vo,	6	00
" Notes and Examples in Mechanics..............8vo,	2	00
Crandall's Earthwork Tables...........................8vo,	1	50
" The Transition Curve..............16mo, morocco,	1	50
* Dredge's Penn. Railroad Construction, etc. Large 4to, half morocco,	20	00
* Drinker's Tunnelling....................4to, half morocco,	25	00
Eissler's Explosives—Nitroglycerine and Dynamite........8vo,	4	00
Folwell's Sewerage....................................8vo,	3	00
Fowler's Coffer-dam Process for Piers...................8vo,	2	50
Gerhard's Sanitary House Inspection...................12mo,	1	00
Godwin's Railroad Engineer's Field-book......16mo, morocco,	2	50
Gore's Elements of Geodesy......... ....8vo,	2	50
Howard's Transition Curve Field-book.........16mo, morocco,	1	50
Howe's Retaining Walls (New Edition.).................12mo,	1	25
Hudson's Excavation Tables. Vol. II................. 8vo,	1	00
Hutton's Mechanical Engineering of Power Plants........8vo,	5	00
Johnson's Materials of Construction...............Large 8vo,	6	00
" Stadia Reduction Diagram..Sheet, 22½ × 28½ inches,		50
" Theory and Practice of Surveying.........Small 8vo,	4	00
Kent's Mechanical Engineer's Pocket-book.....16mo, morocco,	5	00
Kiersted's Sewage Disposal..... ....12mo,	1	25
Mahan's Civil Engineering. (Wood.)....................8vo,	5	00
Merriman and Brook's Handbook for Surveyors....16mo, mor.,	2	00
Merriman's Geodetic Surveying.........................8vo,	2	00
" Retaining Walls and Masonry Dams..........8vo,	2	00
" Sanitary Engineering......................8vo,	2	00
Nagle's Manual for Railroad Engineers........16mo, morocco,	3	00
Ogden's Sewer Design. 12mo,	2	00
Patton's Civil Engineering..................8vo, half morocco,	7	50

Patton's Foundations..................................8vo,	$5 00
Pratt and Alden's Street-railway Road-beds..............8vo,	2 00
Rockwell's Roads and Pavements in France............12mo,	1 25
Searles's Field Engineering...................16mo, morocco,	3 00
" Railroad Spiral.....................16mo, morocco,	1 50
Siebert and Biggin's Modern Stone Cutting and Masonry...8vo,	1 50
Smart's Engineering Laboratory Practice...............12mo,	2 50
Smith's Wire Manufacture and Uses...............Small 4to,	3 00
Spalding's Roads and Pavements......................12mo,	2 00
" Hydraulic Cement........................12mo,	2 00
Taylor's Prismoidal Formulas and Earthwork............8vo,	1 50
Thurston's Materials of Construction 8vo,	5 00
* Trautwine's Civil Engineer's Pocket-book....16mo, morocco,	5 00
* " Cross-section............................Sheet,	25
* " Excavations and Embankments............8vo,	2 00
* " Laying Out Curves............12mo, morocco,	2 50
Waddell's De Pontibus (A Pocket-book for Bridge Engineers). 16mo, morocco,	3 00
Wait's Engineering and Architectural Jurisprudence.......8vo,	6 00
Sheep,	6 50
" Law of Field Operation in Engineering, etc........8vo.	
Warren's Stereotomy—Stone-cutting.....................8vo,	2 50
*Webb's Engineering Instruments............16mo, morocco,	50
" " " New Edition..............	1 25
Wegmann's Construction of Masonry Dams..............4to,	5 00
Wellington's Location of Railways...............Small 8vo,	5 00
Wheeler's Civil Engineering............................8vo,	4 00
Wolff's Windmill as a Prime Mover.....................8vo,	3 00

HYDRAULICS.

WATER-WHEELS—WINDMILLS—SERVICE PIPE—DRAINAGE, ETC.

(*See also* ENGINEERING, p. 7.)

Bazin's Experiments upon the Contraction of the Liquid Vein. (Trautwine.)...................................8vo,	2 00
Bovey's Treatise on Hydraulics........................8vo,	4 00
Coffin's Graphical Solution of Hydraulic Problems.......12mo,	2 50
Ferrel's Treatise on the Winds, Cyclones, and Tornadoes...8vo,	4 00
Fuertes's Water and Public Health....................12mo,	1 50
Ganguillet & Kutter's Flow of Water. (Hering & Trautwine) 8vo,	4 00
Hazen's Filtration of Public Water Supply...............8vo,	2 00
Herschel's 115 Experiments...........................8vo,	2 00

Kiersted's Sewage Disposal............................12mo,	$1	25
Mason's Water Supply..................................8vo,	5	00
" Examination of Water. 12mo,	1	25
Merriman's Treatise on Hydraulics....................8vo,	4	00
Nichols's Water Supply (Chemical and Sanitary)..........8vo,	2	50
Wegmann's Water Supply of the City of New York.......4to,	10	00
Weisbach's Hydraulics. (Du Bois.)....................8vo,	5	00
Whipple's Microscopy of Drinking Water 8vo,	3	50
Wilson's Irrigation Engineering.... ....8vo,	4	00
" Hydraulic and Placer Mining.................12mo,	2	00
Wolff's Windmill as a Prime Mover......................8vo,	3	00
Wood's Theory of Turbines.... ....8vo,	2	50

MANUFACTURES.

BOILERS—EXPLOSIVES—IRON—STEEL—SUGAR—WOOLLENS, ETC.

Allen's Tables for Iron Analysis.........................8vo,	3	00
Beaumont's Woollen and Worsted Manufacture.........12mo,	1	50
Bolland's Encyclopædia of Founding Terms............12mo,	3	00
" The Iron Founder..........................12mo,	2	50
" " " " Supplement...............12mo,	2	50
Bouvier's Handbook on Oil Painting....................12mo,	2	00
Eissler's Explosives, Nitroglycerine and Dynamite.........8vo,	4	00
Fodr's Boiler Making for Boiler Makers.................18mo,	1	00
Metcalfe's Cost of Manufactures........................8vo,	5	00
Metcalf's Steel—A Manual for Steel Users..............12mo,	2	00
*Reisig's Guide to Piece Dyeing........................8vo,	25	00
Spencer's Sugar Manufacturer's Handbook ....16mo, morocco,	2	00
" Handbook for Chemists of Beet Sugar Houses. 16mo, morocco,	3	00
Thurston's Manual of Steam Boilers..................... 8vo,	5	00
Walke's Lectures on Explosives.........................8vo.	4	00
West's American Foundry Practice....................12mo,	2	50
" Moulder's Text-book 12mo,	2	50
Wiechmann's Sugar Analysis..................... Small 8vo,	2	50
Woodbury's Fire Protection of Mills.....................8vo,	2	50

MATERIALS OF ENGINEERING.

STRENGTH—ELASTICITY—RESISTANCE, ETC.

(*See also* ENGINEERING, p. 7.)

Baker's Masonry Construction...........................8vo,	5	00
Beardslee and Kent's Strength of Wrought Iron..........8vo,	1	50
Bovey's Strength of Materials..........................8vo,	7	50
Burr's Elasticity and Resistance of Materials.............8vo,	5	00
Byrne's Highway Construction.........................8vo,	5	00

Church's Mechanics of Engineering—Solids and Fluids.....8vo,	$6 00	
Du Bois's Stresses in Framed Structures............Small 4to,	10 00	
Johnson's Materials of Construction......................8vo,	6 00	
Lanza's Applied Mechanics. 8vo,	7 50	
Martens's Materials. (Henning.)..........8vo. (*In the press.*)		
Merrill's Stones for Building and Decoration................8vo,	5 00	
Merriman's Mechanics of Materials........................8vo,	4 00	
" Strength of Materials.......................12mo,	1 00	
Patton's Treatise on Foundations..........................8vo,	5 00	
Rockwell's Roads and Pavements in France..............12mo,	1 25	
Spalding's Roads and Pavements........................12mo,	2 00	
Thurston's Materials of Construction......................8vo,	5 00	
" Materials of Engineering..............3 vols., 8vo,	8 00	
Vol. I., Non-metallic 8vo,	2 00	
Vol. II., Iron and Steel..........................8vo,	3 50	
Vol. III., Alloys, Brasses, and Bronzes............8vo,	2 50	
Wood's Resistance of Materials...........................8vo,	2 00	

MATHEMATICS.

CALCULUS—GEOMETRY—TRIGONOMETRY, ETC.

Baker's Elliptic Functions..............................8vo,	1 50
Ballard's Pyramid Problem............................8vo,	1 50
Barnard's Pyramid Problem...........................8vo,	1 50
*Bass's Differential Calculus.........................12mo,	4 00
Briggs's Plane Analytical Geometry....................12mo,	1 00
Chapman's Theory of Equations......................12mo,	1 50
Compton's Logarithmic Computations..................12mo,	1 50
Davis's Introduction to the Logic of Algebra..............8vo,	1 50
Halsted's Elements of Geometry........................8vo,	1 75
" Synthetic Geometry.........................8vo,	1 50
Johnson's Curve Tracing..............................12mo,	1 00
" Differential Equations—Ordinary and Partial. Small 8vo,	3 50
" Integral Calculus........................12mo,	1 50
" " " Unabridged. Small 8vo. (*In the press.*)	
" Least Squares..........................12mo,	1 50
*Ludlow's Logarithmic and Other Tables. (Bass.)........8vo,	2 00
* " Trigonometry with Tables. (Bass.)...........8vo,	3 00
*Mahan's Descriptive Geometry (Stone Cutting) 8vo,	1 50
Merriman and Woodward's Higher Mathematics...........8vo,	5 00
Merriman's Method of Least Squares....................8vo,	2 00
Parker's Quadrature of the Circle 8vo,	2 50
Rice and Johnson's Differential and Integral Calculus, 2 vols. in 1, small 8vo,	2 50

Rice and Johnson's Differential Calculus..........Small 8vo,	$3	00
" Abridgment of Differential Calculus.		
Small 8vo,	1	50
Totten's Metrology..8vo,	2	50
Warren's Descriptive Geometry..................2 vols., 8vo,	3	50
" Drafting Instruments........................12mo,	1	25
" Free-hand Drawing...........................12mo,	1	00
" Higher Linear Perspective....................8vo,	3	50
" Linear Perspective...........................12mo,	1	00
" Primary Geometry...........................12mo,		75
" Plane Problems..............................12mo,	1	25
" Problems and Theorems.......................8vo,	2	50
" Projection Drawing..........................12mo,	1	50
Wood's Co-ordinate Geometry...........................8vo,	2	00
" Trigonometry...................................12mo,	1	00
Woolf's Descriptive Geometry.....................Large 8vo,	3	00

MECHANICS—MACHINERY.

TEXT-BOOKS AND PRACTICAL WORKS.

(*See also* ENGINEERING, p. 7.)

Baldwin's Steam Heating for Buildings................12mo,	2	50
Benjamin's Wrinkles and Recipes......................12mo,	2	00
Chordal's Letters to Mechanics.......................12mo,	2	00
Church's Mechanics of Engineering.....................8vo,	6	00
" Notes and Examples in Mechanics............8vo,	2	00
Crehore's Mechanics of the Girder.....................8vo,	5	00
Cromwell's Belts and Pulleys........................12mo,	1	50
" Toothed Gearing.............................12mo,	1	50
Compton's First Lessons in Metal Working............12mo,	1	50
Compton and De Groodt's Speed Lathe................12mo,	1	50
Dana's Elementary Mechanics........................12mo,	1	50
Dingey's Machinery Pattern Making...................12mo,	2	00
Dredge's Trans. Exhibits Building, World Exposition.		
Large 4to, half morocco,	10	00
Du Bois's Mechanics. Vol. I., Kinematics 8vo,	3	50
" " Vol. II., Statics.. ....8vo,	4	00
" " Vol. III., Kinetics........ ....8vo,	3	50
Fitzgerald's Boston Machinist......................18mo,	1	00
Flather's Dynamometers.............................12mo,	2	00
" Rope Driving................................12mo,	2	00
Hall's Car Lubrication..............................12mo,	1	00
Holly's Saw Filing 18mo,		75
Johnson's Theoretical Mechanics. An Elementary Treatise		
(*In the press.*)		
Jones's Machine Design. Part I., Kinematics...........8vo,	1	50

Jones's Machine Design. Part II., Strength and Proportion of Machine Parts............8vo,	$3	00
Lanza's Applied Mechanics............8vo,	7	50
MacCord's Kinematics............8vo,	5	00
Merriman's Mechanics of Materials............8vo,	4	00
Metcalfe's Cost of Manufactures............8vo,	5	00
*Michie's Analytical Mechanics............8vo,	4	00
Richards's Compressed Air............12mo,	1	50
Robinson's Principles of Mechanism............8vo,	3	00
Smith's Press-working of Metals............8vo,	3	00
Thurston's Friction and Lost Work............8vo,	3	00
" The Animal as a Machine............12mo,	1	00
Warren's Machine Construction............2 vols., 8vo,	7	50
Weisbach's Hydraulics and Hydraulic Motors. (Du Bois.)..8vo,	5	00
" Mechanics of Engineering. Vol. III., Part I., Sec. I. (Klein.)............8vo,	5	00
Weisbach's Mechanics of Engineering. Vol. III., Part I., Sec. II. (Klein.)............8vo,	5	00
Weisbach's Steam Engines. (Du Bois.)............8vo,	5	00
Wood's Analytical Mechanics............8vo,	3	00
" Elementary Mechanics............12mo,	1	25
" " " Supplement and Key............12mo,	1	25

METALLURGY.

Iron—Gold—Silver—Alloys, Etc.

Allen's Tables for Iron Analysis............8vo,	3	00
Egleston's Gold and Mercury............Large 8vo,	7	50
" Metallurgy of Silver............Large 8vo,	7	50
* Kerl's Metallurgy—Copper and Iron............8vo,	15	00
* " " Steel, Fuel, etc............8vo,	15	00
Kunhardt's Ore Dressing in Europe............8vo,	1	50
Metcalf's Steel—A Manual for Steel Users............12mo,	2	00
O'Driscoll's Treatment of Gold Ores............8vo,	2	00
Thurston's Iron and Steel............8vo,	3	50
" Alloys............8vo,	2	50
Wilson's Cyanide Processes............12mo,	1	50

MINERALOGY AND MINING.

Mine Accidents—Ventilation—Ore Dressing, Etc.

Barringer's Minerals of Commercial Value....Oblong morocco,	2	50
Beard's Ventilation of Mines............12mo,	2	50
Boyd's Resources of South Western Virginia............8vo,	3	00
" Map of South Western Virginia......Pocket-book form,	2	00

Brush and Penfield's Determinative Mineralogy. New Ed. 8vo,	$4	00
Chester's Catalogue of Minerals..........................8vo,	1	25
" " " " ....Paper,		50
" Dictionary of the Names of Minerals............8vo,	3	00
Dana's American Localities of Minerals............Large 8vo,	1	00
" Descriptive Mineralogy. (E. S.)....Large half morocco,	12	50
" Mineralogy and Petrography. (J. D.)............12mo,	2	00
" Minerals and How to Study Them. (E. S.).....12mo,	1	50
" Text-book of Mineralogy. (E. S.)...New Edition. 8vo,	4	00
* Drinker's Tunnelling, Explosives, Compounds, and Rock Drills. 4to, half morocco,	25	00
Egleston's Catalogue of Minerals and Synonyms...........8vo,	2	50
Eissler's Explosives—Nitroglycerine and Dynamite........8vo,	4	00
Hussak's Rock-forming Minerals. (Smith.).........Small 8vo,	2	00
Ihlseng's Manual of Mining............................8vo,	4	00
Kunhardt's Ore Dressing in Europe.....................8vo,	1	50
O'Driscoll's Treatment of Gold Ores....................8vo,	2	00
* Penfield's Record of Mineral Tests..........Paper, 8vo,		50
Rosenbusch's Microscopical Physiography of Minerals and Rocks. (Iddings.)..............................8vo,	5	00
Sawyer's Accidents in Mines......................Large 8vo,	7	00
Stockbridge's Rocks and Soils..........................8vo,	2	50
Walke's Lectures on Explosives.........................8vo,	4	00
Williams's Lithology...................................8vo,	3	00
Wilson's Mine Ventilation.............................12mo,	1	25
" Hydraulic and Placer Mining...............12mo,	2	50

STEAM AND ELECTRICAL ENGINES, BOILERS, Etc.

STATIONARY—MARINE—LOCOMOTIVE—GAS ENGINES, ETC.

(*See also* ENGINEERING, p. 7.)

Baldwin's Steam Heating for Buildings................12mo,	2	50
Clerk's Gas Engine.............................Small 8vo,	4	00
Ford's Boiler Making for Boiler Makers................18mo,	1	00
Hemenway's Indicator Practice........................12mo,	2	00
Hoadley's Warm-blast Furnace..........................8vo,	1	50
Kneass's Practice and Theory of the Injector..........8vo,	1	50
MacCord's Slide Valve.................................8vo,	2	00
Meyer's Modern Locomotive Construction................4to,	10	00
Peabody and Miller's Steam-boilers....................8vo,	4	00
Peabody's Tables of Saturated Steam...................8vo,	1	00
" Thermodynamics of the Steam Engine........ 8vo,	5	00
" Valve Gears for the Steam Engine...........8vo,	2	50
Pray's Twenty Years with the Indicator..........Large 8vo,	2	50
Pupin and Osterberg's Thermodynamics................12mo,	1	25

Reagan's Steam and Electric Locomotives..........12mo,	$2 00
Röntgen's Thermodynamics. (Du Bois.)..............8vo,	5 00
Sinclair's Locomotive Running.....................12mo,	2 00
Snow's Steam-boiler Practice........8vo. (*In the press.*)	
Thurston's Boiler Explosions......................12mo,	1 50
" Engine and Boiler Trials....................8vo,	5 00
" Manual of the Steam Engine. Part I., Structure and Theory.............................8vo,	6 00
" Manual of the Steam Engine. Part II., Design, Construction, and Operation..............8vo,	6 00
2 parts,	10 00
Thurston's Philosophy of the Steam Engine..........12mo,	75
" Reflection on the Motive Power of Heat. (Carnot.) 12mo,	1 50
" Stationary Steam Engines....................8vo,	2 50
" Steam-boiler Construction and Operation.......8vo,	5 00
Spangler's Valve Gears.................................8vo,	2 50
Weisbach's Steam Engine. (Du Bois.)..................8vo,	5 00
Whitham's Constructive Steam Engineering.............8vo,	6 00
" Steam-engine Design........................8vo,	5 00
Wilson's Steam Boilers. (Flather.)....................12mo,	2 50
Wood's Thermodynamics, Heat Motors, etc.............8vo,	4 00

TABLES, WEIGHTS, AND MEASURES.

FOR ACTUARIES, CHEMISTS, ENGINEERS, MECHANICS—METRIC TABLES, ETC.

Adriance's Laboratory Calculations...................12mo,	1 25
Allen's Tables for Iron Analysis........................8vo,	3 00
Bixby's Graphical Computing Tables..................Sheet,	25
Compton's Logarithms..............................12mo,	1 50
Crandall's Railway and Earthwork Tables..............8vo,	1 50
Egleston's Weights and Measures.....................18mo,	75
Fisher's Table of Cubic Yards....................Cardboard,	25
Hudson's Excavation Tables. Vol. II..................8vo,	1 00
Johnson's Stadia and Earthwork Tables................8vo,	1 25
Ludlow's Logarithmic and Other Tables. (Bass.).......12mo,	2 00
Totten's Metrology..................................8vo,	2 50

VENTILATION.

STEAM HEATING—HOUSE INSPECTION—MINE VENTILATION.

Baldwin's Steam Heating............................12mo,	2 50
Beard's Ventilation of Mines.........................12mo,	2 50
Carpenter's Heating and Ventilating of Buildings.........8vo,	3 00
Gerhard's Sanitary House Inspection..................12mo,	1 00
Reid's Ventilation of American Dwellings..............12mo,	1 50
Wilson's Mine Ventilation...........................12mo,	1 25

MISCELLANEOUS PUBLICATIONS.

Alcott's Gems, Sentiment, Language............Gilt edges,	$5 00
Bailey's The New Tale of a Tub..........................8vo,	75
Ballard's Solution of the Pyramid Problem.............8vo,	1 50
Barnard's The Metrological System of the Great Pyramid..8vo,	1 50
Davis's Elements of Law..................................8vo,	2 00
Emmon's Geological Guide-book of the Rocky Mountains..8vo,	1 50
Ferrel's Treatise on the Winds..........................8vo,	4 00
Haines's Addresses Delivered before the Am. Ry. Assn...12mo,	2 50
Mott's The Fallacy of the Present Theory of Sound..Sq. 16mo,	1 00
Perkins's Cornell University....................Oblong 4to,	1 50
Ricketts's History of Rensselaer Polytechnic Institute.....8vo,	3 00
Rotherham's The New Testament Critically Emphasized. 12mo,	1 50
" The Emphasized New Test. A new translation. Large 8vo,	2 00
Totten's An Important Question in Metrology............8vo,	2 50
Whitehouse's Lake Mœris..............................Paper,	25
* Wiley's Yosemite, Alaska, and Yellowstone............4to,	3 00

HEBREW AND CHALDEE TEXT-BOOKS.
FOR SCHOOLS AND THEOLOGICAL SEMINARIES.

Gesenius's Hebrew and Chaldee Lexicon to Old Testament. (Tregelles.)..................Small 4to, half morocco,	5 00
Green's Elementary Hebrew Grammar................12mo,	1 25
" Grammar of the Hebrew Language (New Edition).8vo,	3 00
" Hebrew Chrestomathy........................8vo,	2 00
Letteris's Hebrew Bible (Massoretic Notes in English). 8vo, arabesque,	2 25

MEDICAL.

Bull's Maternal Management in Health and Disease.......12mo,	1 00
Hammarsten's Physiological Chemistry. (Mandel.)........8vo,	4 00
Mott's Composition, Digestibility, and Nutritive Value of Food. Large mounted chart,	1 25
Ruddiman's Incompatibilities in Prescriptions............8vo,	2 00
Steel's Treatise on the Diseases of the Ox.... ....8vo,	6 00
" Treatise on the Diseases of the Dog..............8vo,	3 50
Woodhull's Military Hygiene.........................16mo,	1 50
Worcester's Small Hospitals—Establishment and Maintenance, including Atkinson's Suggestions for Hospital Architecture.................................12mo,	1 25

www.ingramcontent.com/pod-product-compliance
Lightning Source LLC
Chambersburg PA
CBHW030302170426
43202CB00009B/839